Holley Carburetors

HOW TO REBUILD

Mike Mavrigian

CarTech®

CarTech®

CarTech®, Inc.
6118 Main Street
North Branch, MN 55056
Phone: 651-277-1200 or 800-551-4754
Fax: 651-277-1203
www.cartechbooks.com

Edit by Paul Johnson
Layout by Monica Seiberlich

ISBN 978-1-61325-198-0
Item No. SA330

Library of Congress Cataloging-in-Publication Data

Names: Mavrigian, Mike, author.
Title: Holley carburetors : how to rebuild / Mike Mavrigian.
Description: Forest Lake, MN : CarTech, [2016]
Identifiers: LCCN 2015014706 | ISBN 9781613251980
Subjects: LCSH: Holley carburetors. | Holley carburetors–Maintenance and repair.
Classification: LCC TL212 .M38 2016 | DDC 629.25/330288–dc23
LC record available at http://lccn.loc.gov/2015014706

Written, edited, and designed in the U.S.A.
Printed in China
10 9 8 7 6

Title Page:
When it comes time to rebuild your Holley carb, tools and parts selection is important. This stand allows pivoting and rotating of the carb a full 360-degrees.

Back Cover Photos

Top:
One of the first steps in disassembly is to remove the float bowl. Four screws hold the float bowl to the carb body. Once the screws have been removed, a light tap with a plastic mallet usually frees the float bowl from the metering plate.

Middle Left:
Install the main jets to the metering block using a dedicated jet driver instead of a conventional screwdriver. A Holley jet tool safely captures the jet and eliminates accidental slippage or burring of the jet's drive slot.

Middle Right:
On the metering block, you find the idle-mixture screw. Notice the cork-rubber gasket in the hole recess. Use a small pick to remove this gasket. Always plan to replace this gasket because it can be easily damaged.

Bottom:
When assembling the carb, slide the fuel transfer tube on the float bowl. Slide the fuel bowl into position while allowing the fuel tube to slide into its rubber seal. When fully installed, the bulge at each end of the tube should be seated against its rubber seal.

DISTRIBUTION BY:

Europe
PGUK
63 Hatton Garden
London EC1N 8LE, England
Phone: 020 7061 1980 • Fax: 020 7242 3725
www.pguk.co.uk

Australia
Renniks Publications Ltd.
3/37-39 Green Street
Banksmeadow, NSW 2109, Australia
Phone: 2 9695 7055 • Fax: 2 9695 7355
www.renniks.com

Canada
Login Canada
300 Saulteaux Crescent
Winnipeg, MB, R3J-3T2 Canada
Phone: 800 665 1148 • Fax: 800 665 0103
www.lb.ca

CONTENTS

ACKNOWLEDGMENTS

Thanks to the following for their assistance in producing this book: Bill Tichenor, Holley; Dave Monyhan, Goodson Tools; Scott Gressman, Gressman Powersports; Brian Carruth, Trick Flow Specialties; Scott Koffel, Koffel's Place; Tony Lombardi, Ross Racing Engines; Trey, AED; Diro Ahart, CRC; Bob Harris, Lisle Corp.; Robert Holmes, Mac Tools; Thor Schroeder, Moroso; Alan Rebescher, Summit Racing; Willy, Willy's Carbs; Lista International; and Jody Holtry, Medina Mountain Motors.

INTRODUCTION

During the 1950s and 1960s, the Holley 4-barrel carburetor established itself as the clear leader in high-performance operation because it provided precise fuel metering in low-, mid-, and high-RPM operation. Also, it was a modular design that was easy to tune and had a wide range of parts. The Holley 4150-series, or Double Pumper, carburetor distinguished itself during the muscle car era because it had an accelerator arm pump on both float bowls, so if the carburetor were used for aggressive acceleration or wide-open throttle, it would deliver all the needed fuel into the engine when called upon.

Although the basic architecture of the Holley 4-barrel carburetor has remained largely the same for 60 years, Holley has offered several different series of carburetors, such as the 4150, 4160, Avenger, and Dominator. And of course, many different carburetor models exist within these series. In this book, I review Holley carb styles, models, features, and sizes, so you can accurately identify your particular carburetor and its parts.

Many classic cars, muscle cars, hot rods, street rods, and competition cars are equipped with Holley carburetors. If you have an OEM Holley carburetor, it is most likely 30 years or older. Over time, gaskets deteriorate and bushings wear and possibly leak, metering blocks and jets are exposed to contaminants and become plugged, jet needles and jets wear from use and no longer meter properly. If you have an aftermarket carburetor, it may be newer but a lot of usage may require a rebuild. Or possibly, you want to substantially change the setup of your carburetor. Whatever your situation, if you have never worked on a carburetor before, rebuilding one can seem daunting.

The purpose of this book is to take the mystery out of the carburetor rebuild process, and to show you how to do it in straightforward, clear steps. This includes how to disassemble, select parts, install components, and confidently reassemble a carburetor, so your engine and vehicle performs at its best. In the pages of this book, I have documented the entire Holley carburetor rebuild procedure so you disassemble, evaluate, replace worn/defective parts, reassemble, and perform basic calibration. The step-by-step photos illustrate how to perform each step, and the captions explain the tools and techniques required to perform the procedure.

Before you dive into rebuilding a Holley carburetor, it's important to understand how it operates. Carburetors are relatively complex devices, and they have a number of circuits, venturis, orifices, and other features. You need to understand how a carburetor and each component functions so you can properly rebuild and tune it. I have included a comprehensive parts guide that provides the available parts for a particular model, so you can select the correct jets, needles, emulsion tubes, and other parts for optimal performance. I also help you select a vacuum or mechanical secondary carburetor for your particular engine and application. Holleys can be set up and calibrated for almost any engine, vehicle, and application.

If you have built a modified engine or if you're buying a new carburetor for a particular engine, you need to select the correct CFM (cubic flow per minute) size. Feed too much fuel to the engine and it will sputter and stumble or, in extreme cases, not start. Supply too little fuel and the engine runs out of power on the top end. The instruction and mathematical formulas allow you to select the right carburetor.

Holley 4-barrel carburetors have a number of jets, needle jets, main jets, and power jets; all of the parts must be tuned as a well-integrated system if you want to get maximum fuel efficiency and performance. This book does not delve into performance modifications. But if you're looking to setup your Holley Dominator, which is installed on a big-block Chevy, and you plan on drag racing, I suggest you refer to *How to Super Tune and Modify Holley Carburetors* by David Vizard. In that book, Vizard uses 50-plus years of engine calibration experience and knowledge so you can tune your carburetor for maximum performance for almost any application, high-performance street, drag race, road race, and almost every other form of competition. He also explains how to calibrate the carburetor for almost any engine setup.

With this book in hand, you will be better able to understand your Holley's functions and how to perform do-it-yourself repairs and rebuilding to obtain a fresh, like-new carb, in terms of on-engine function. If you've never worked on a Holley carburetor before, or if you have dabbled but have questions regarding specific components and disassembly/assembly operations, this book is designed to answer all of your questions. I've also included information that is helpful when choosing a carb for a specific application, in terms of features and CFM size.

Welcome to the world of Holley carburetors.

WHAT IS A WORKBENCH® BOOK?

This Workbench® Series book is the only book of its kind on the market. No other book offers the same combination of detailed hands-on information and close-up photographs to illustrate rebuilding and modifying. Rest assured, you have purchased an indispensable companion that will expertly guide you, one step at a time, through each important stage of the rebuilding process. This book is packed with real-world techniques and practical tips for expertly performing rebuild procedures, not vague instructions or unnecessary processes. At-home mechanics or enthusiast builders strive for professional results, and the instruction in our Workbench® Series books help you realize pro-caliber results. Hundreds of photos guide you through the entire process from start to finish, with informative captions containing comprehensive instructions for every step of the process.

The step-by-step photo procedures also contain many additional photos that show how to install high-performance components, modify stock components for special applications, or even call attention to assembly steps that are critical to proper operation or safety. These are labeled with unique icons. These symbols represent an idea, and photos marked with the icons contain important, specialized information.

Here are some of the icons found in Workbench® books:

Important!

Calls special attention to a step or procedure, so that the procedure is correctly performed. This prevents damage to a vehicle, system, or component.

Save Money

Illustrates a method or alternate method of performing a rebuild step that will save money but still give acceptable results.

Torque Fasteners

Illustrates a fastener that must be properly tightened with a torque wrench at this point in the rebuild. The torque specs are usually provided in the step.

Special Tool

Illustrates the use of a special tool that may be required or can make the job easier (caption with photo explains further).

Performance Tip

Indicates a procedure or modification that can improve performance. The step most often applies to high-performance or racing engines.

Critical Inspection

Indicates that a component must be inspected to ensure proper operation of the engine.

Precision Measurement

Illustrates a precision measurement or adjustment that is required at this point in the rebuild.

Professional Mechanic Tip

Illustrates a step in the rebuild that non-professionals may not know. It may illustrate a shortcut or a trick to improve reliability, prevent component damage, etc.

Documentation Required

Illustrates a point in the rebuild where the reader should write down a particular measurement, size, part number, etc. for later reference or photograph a part, area, or system of the vehicle for future reference.

Tech Tip

Tech Tips provide brief coverage of important subject matter that doesn't naturally fall into the text or step-by-step procedures of a chapter. Tech Tips contain valuable hints, important info, or outstanding products that professionals have discovered after years of work. These will add to your understanding of the process, and help you get the most power, economy, and reliability from your engine.

HOLLEY HISTORY, MODELS AND SERIAL NUMBERS

The 1950s saw the introduction of the Holley Series 4150 4-barrel on the 1957 Thunderbird. It was the beginning of the modular Holley 4-barrel, as we know it today. It was the first true performance carburetor and it became standard equipment on many high-performance automobiles.

The 1960s were huge for the hot rod industry and Holley; the 4150 became original equipment on many iconic Detroit factory muscle cars, powering such notable beasts as the popular Z28 Camaros, big-block Chevelles, Boss Mustangs, and Shelby Cobras. This era also saw the introduction of the Holley three-deuce multi-carb setups on 427 (Tri Power) Corvettes and 440 (Six Pack) Mopars. The now-famous Holley Double Pumper was also born in the 1960s. According to Holley, the now-world-renowned Dominator, which is a Double Pumper, was developed specifically for NASCAR racing and made its debut in 1969.

The 1970s saw Holley's continued dominance in racing with nearly every factory NHRA Super Stock/Pro Stock racer running Holleys. That hasn't changed; Holley carbs have powered more drag racers than all other carbs combined. This era also witnessed the introduction of

This carb (PN 1850) is an example from the 4160 series. It is a 4-barrel carb that has a single feed, a primary metering block with jets, one accelerator pump at the primary side, and a metering plate at the secondary side.

Holley aluminum intake manifolds, including the once-popular Z-Series developed in conjunction with Zora Arkus-Duntov.

The 1980s marked Holley's entrance into the fuel-injection market for which original-equipment EFI components and analog Pro-Jection retrofit fuel-injection systems for carbureted cars were introduced.

In the 1990s Holley continued its new product introductions. The wildly popular HP Pro series of race-ready carburetors was introduced and has become the standard in racing. SysteMAX engine kits were introduced with matched cylinder heads, intakes, and cams. Also in the 1990s, the Dominator evolved into the HP Dominator, huge billet electric fuel pumps were introduced, and retrofit EFI kits evolved into digital Pro-Jection 4D and 4Di.

Beginning in the 2000s, Holley introduced the popular Street Avenger, Truck Avenger, Street HP, and Ultra HP carburetors as well as billet mechanical fuel pumps and high-flow billet electric pumps. Although not covered in this book,

Holley's EFI systems have seen an enormous amount of change with the introduction of Avenger EFI, HP EFI, and Dominator EFI.

Holley ID System

Two basic numbers are attributed to each Holley carburetor: the model/series number and the list/part number. The series number indicates the general type, or series; for instance, 4150, 4160, or 4500.

The 4150 series is the original 4-barrel modular design, which features metering blocks and replaceable jets on both the primary and secondary sides. The 4150 series is available with either mechanical or vacuum secondary operation.

The 4160 series is a slightly less-expensive variant of the 4150, featuring replaceable primary jets, but with a thin metering plate on the secondary side and no replaceable secondary jets.

The 4500 series is the Dominator big-CFM racing series, with a larger main body and a unique secondary actuation linkage design.

The model number is not stamped on the carburetor, but it's easy to identify by simply examining the carburetor. If the secondary side has a metering block with jets, it's a 4150; if not, it's a 4160. The Dominator is easily identifiable by its size and the shape of the throttle bores.

For all practical purposes, the "list" number is the part number of the carburetor; it is the top number stamped on the carb. The list number is also stamped on the choke housing/air horn of all Holley carburetors that have a choke housing. Keep in mind that the Dominator series and some other race carbs, such as the Ultra HP series, do not have a choke

The list number (also called the part number) appears on the face of the choke housing on carbs that have chokes. The word "LIST" or the letter "L" sometimes precedes the part number; these were used randomly and have nothing to do with the year of manufacture.

As seen in this example (a 4160-series 600-cfm carb), 1850 is the part number. Following the part number is –2, which indicates that this is the 12th update or change for this carb; this likely means that the dies have been renewed 12 times. For models that have been produced for decades, dies and tooling wear and necessitate updates in the manufacturing process.

You don't need to be concerned with the "dash" number. The bottom number is the build date. The first three numerals represent how many days from the start of a year that a carb was made, and the fourth numeral is the last digit of the year of manufacture. This carb was produced on the 133rd day of 1974, 1984, 1994, or 2004.

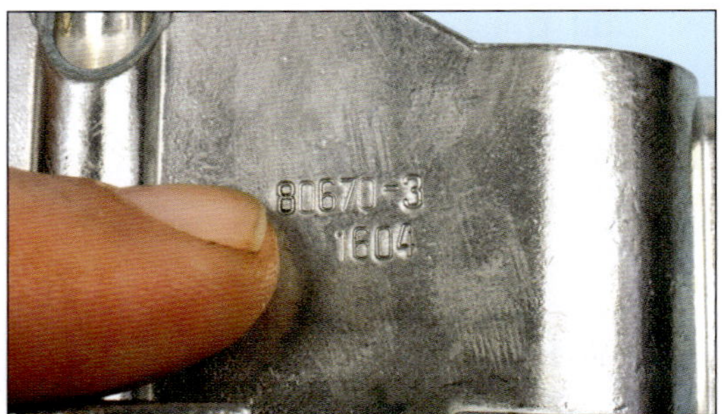

This carburetor's choke housing is stamped 80670-3, which indicates that this is a 670-cfm Street Avenger. The number shown below the part number indicates that it was built on the 160th day of a year ending in 4. Based on when this carb was purchased, this likely means 2014.

This HP carb, because it has no choke housing, carries its identification on the side of the throttle body. It is clearly marked as a 4150. Note the "HP" designation under "4150." Under the word LIST appears the list, or part, number of 82851, which indicates that it's an 850-cfm Street HP carb. The build date of 0471 indicates that it was made on the 47th day of a year ending in 1. Based on when this carb was purchased, it was likely built in 2011.

Street HP carbs have a vibratory polished body as well as smoothly contoured venturi inlets. All Holley carbs can be easily identified; simply use the part number stamped on the carb and search the database on Holley's website.

Carburetors such as the Dominator 4500 series do not have a choke housing, and as a result, the part number and date code are located on the side of the main throttle body.

This example shows a list number (also known as the part number) of 8082-3, indicating that it's a 1,050-cfm HP unit for a single 4-barrel installation. Depending on design, Dominators are avail-able designed for 1 x 4 (one 4-barrel carb) or 2 x 4 (two 4-barrel carbs for a dual setup) mounted to the intake manifold.

Although 4150- and 4160-series carbs do not display the model number, the Dominator identification does include the series number, which in this case is 4500. The build date indicates the 203rd day of a year ending in 4. Since this carb was recently purchased, I assume that it was built in 2004 or 2014.

housing; their list number is stamped on the top of the main body.

The list number *may be* preceded by either the word "LIST" or by the letter "L." The factory used the word "LIST" or the letter "L" randomly. Don't worry about it. Some carbs have it and some don't.

For example, a LIST 3310 is either a 4160 at 750 cfm or a 4150 at 780 cfm, depending on when it was made. Another example is a Street Avenger carb's choke housing/air horn stamped with the part number (e.g. 80670), without "LIST" or "L."

A dash followed by a single- or double-digit number may be found immediately following the part num-ber. This dash number simply indicates that a running change for that partic-ular part number of carb had occurred (perhaps a die was changed, or other evolutionary production change). A –1 indicates the first running change. A –4 indicates the fourth running change, etc. From a selection/purchas-ing standpoint, you don't need to be concerned with this number.

Located underneath the part number is a four-digit number, which indicates the build date. For instance, a build date might be 1954. This indicates that this particular carb was built on the 195th day of a year end-ing in "4" (which could mean 1974, 1984, 1994, or 2004).

Although this may seem con-fusing, dating a Holley carburetor is generally not too difficult. A date code should appear below the list number. Older carburetors have a three-digit code, while carbs made after 1972 used a four-digit code. You can always call Holley's tech hotline and ask a Holley technician to help decipher the date code, but remem-ber, this number doesn't have any particular use for you, so you really don't need to worry about it.

Series 4150 and 4160

Holley 4150s and 4160s are built on similar platforms with a few dis-tinct differences. They both have square bores; the primary and sec-ondary throttle bores are the same size, in a "square" configuration, in contrast to a spread-bore that has smaller primary and larger secondary bores.

The 4150 is a square-bore carb with center-hung floats and dual fuel feed inlets, and replaceable jets in both primary and secondary meter-ing blocks. The 4150 carbs with vacuum-operated secondaries have an accelerator pump and a power valve on the primary bowl, but no accelerator pump or power valve on the secondary bowl.

The 4150 carbs with mechani-cal secondaries are known as Double Pumper carbs; they have the addition of a secondary accelerator pump, and depending on the specific part num-ber, may also have a power valve on the secondary side.

The Series 4160 carbs also have a square-bore pattern but with more basic features. The primary side has a metering block with jets, but the sec-ondary side has a thin metering plate with pre-sized orifices.

Although a metering block has replaceable jets for tuning purposes, a metering plate (used in the second-ary side of 4160 carbs) requires chang-ing the metering plate for secondary "jet" size tuning. Metering plates are available in a range of main-hole and idle-hole configurations. Secondary metering plates are secured to the main body with six 8-32 clutch-head screws, so you should obtain a 5/32-inch clutch-head driver for proper servicing during removal and installation.

A 4160 can be converted to a 4150 by adding a secondary metering block in place of the metering plate and swapping to center–hung-float fuel bowls. A 4150 can be converted to a 4160, although it is not a popu-lar modification.

This side-by-side comparison of fuel-bowl styles makes them easy to identify. At the left is a 4160 (PN 1850) carb equipped with a side-hung float. At the right is a 4150 carb with a center-hung float. The 4150 carb has a metering block at the secondary side; the 4160 has a secondary metering plate. A 4160 is easily converted to the 4150 configuration.

4160 Details

This 4160 (PN 1850) carb has the metering block on the primary side (left) and no metering block at the secondary side (right). The secondary side includes a metering plate with no replaceable jets.

Because the 4160 carb has only a front/primary side fuel inlet, fuel is transferred to the rear bowl via this external fuel transfer tube.

The 4160 carb has a single fuel inlet that accepts a slip-on fuel hose. This banjo-style fitting can easily be changed to a banjo fitting with a −6 AN threaded hose-end connection.

4160 Details *CONTINUED*

Fuel bowls have either side-hung or center-hung floats. This is easy to identify by looking at the shape of the fuel bowl. If it is rectangular without a "V" protrusion on the casting, the bowl has a side-hung float, such as this example (PN 1850).

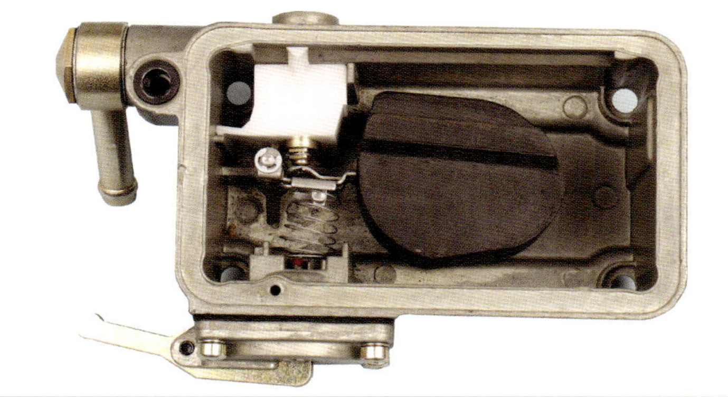

This fuel bowl, removed from a 4160-series carb (PN 1850) reveals a side-hung float, meaning that the float is hinged at one side. A side-hung float may be more susceptible to fuel-level changes during hard turns in contrast to a center-hung float that maintains fuel level better during turns.

With the front bowl removed from this 4160-series carb, you can clearly see the jets near the bottom and the power valve at the lower center of the metering block.

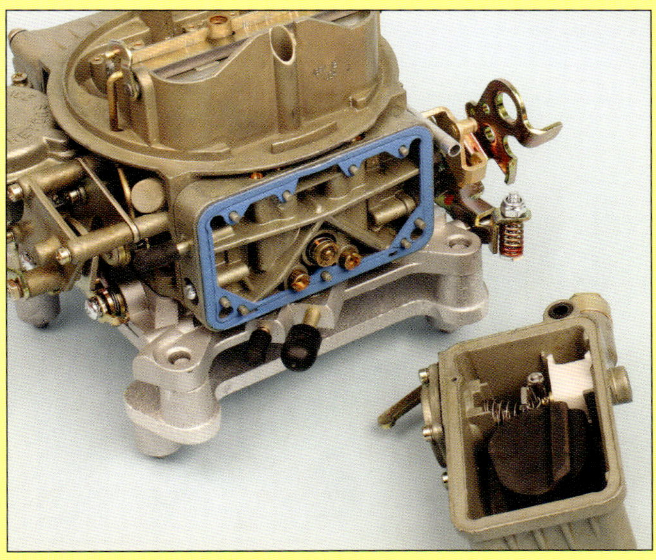

With the front fuel bowl removed from this 4160 carb, the primary metering block is exposed; it is sandwiched between the body and bowl, and is gasketed on each side. The fuel bowl's four screws secure both the metering block and bowl to the carb body.

When the primary fuel bowl is removed, you can see that the metering block is held to the body by the gasket. Small alignment dowels serve to center the block in place.

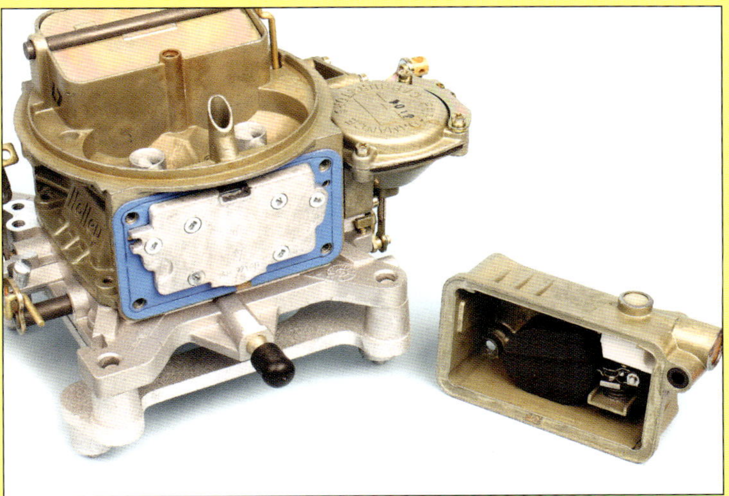

Removing the secondary fuel bowl from this 4160 carb reveals a metering plate, which is secured to the body with six small flat-top screws.

4160 Details *CONTINUED*

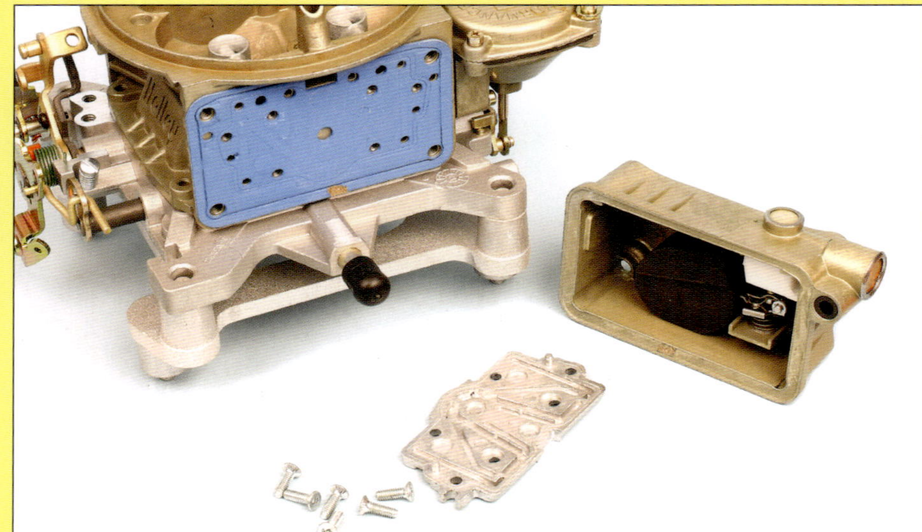

The 4160-series carbs are easily converted to 4150 status: Remove the metering plate and its gasket and then install a metering block with the appropriate gasket and longer fuel-bowl screws.

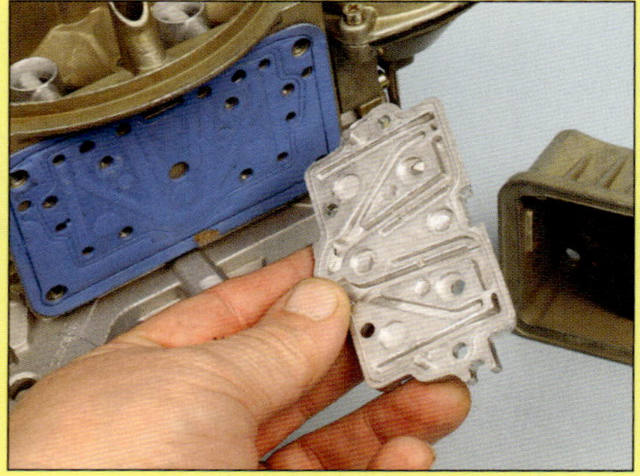

A metering plate provides a "fixed" secondary enrichment. A metering block allows fine-tuning with replaceable jets.

The front and rear fuel bowls, front metering block, and rear metering plate on this 4160 carb have been removed. Note the external fuel transfer tube on this carb (PN 1850).

4150 Details

This is an example of a 4150 carb (PN 4779). It has 750 cfm, double-pumper operation with jet-equipped primary and secondary metering blocks and mechanical choke.

Holley Double Pumper carbs are available in various finishes, including this traditional chromate.

4150 Details CONTINUED

All 4150 carburetors have metering blocks at both the primary and the secondary sides.

All Double Pumper carbs include a dedicated accelerator pump for the rear secondary side.

Double Pumper carbs have fuel bowl level-checking ports. This carb (PN 4779) has the traditional brass-thread plugs that are removed to check fuel bowl level while adjusting float level.

Center-hung fuel bowl floats have needle and seat float adjustment. To make an adjustment, loosen the slotted plug (which is the lock), turn the hex to make the adjustment, and then tighten the lock.

This street Ultra HP has red billet anodized metering plates and baseplate.

Notice the "V" shape on this fuel bowl casting, which clearly indicates the use of a center-hung float.

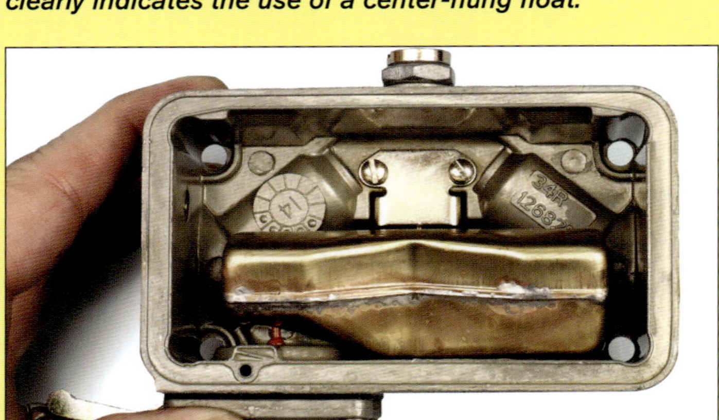

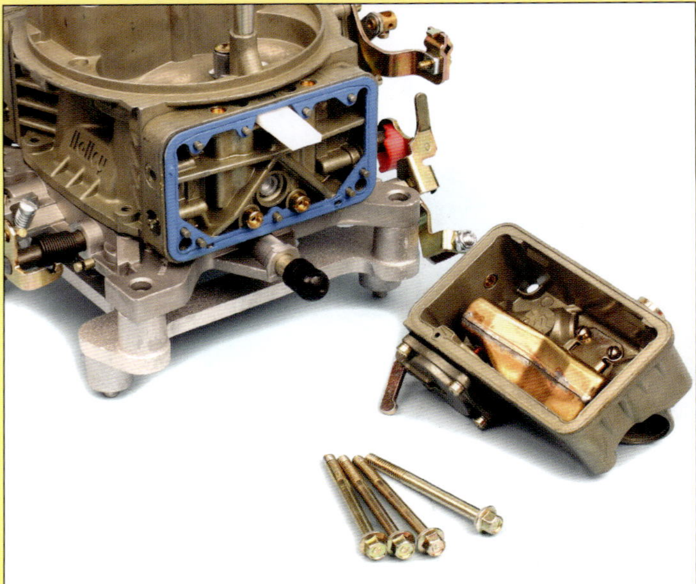

Here, the primary fuel bowl has been removed on a 4150 to reveal the primary metering block.

Center-hung floats pivot at the upper center of the fuel bowl and are less susceptible to fuel-bowl level changes during hard-driving maneuvers.

The jets on the metering block thread in at the front side; the power-valve threads in from the rear of the block. The power valve is rated at a specific vacuum, in inches of mercury. They are usually found only on the primary side but are also installed on the secondary metering block of certain high-performance models.

This is part of the fuel enrichment circuit that opens gradually to full open, depending on engine vacuum. During cruising, the valve is shut. As engine-load demand increases upon acceleration, engine vacuum begins to drop, allowing the power valve to open. Power valves are available in a range of vacuum ratings, which allows you to tailor it according to your needs.

A Holley power valve includes a four-sided head that can be serviced with a 1-inch wrench, but the flats are fairly thin. A dedicated power-valve wrench is highly recommended.

This white plastic piece is commonly referred to as a "whistle." It aids in preventing fuel from spilling out of the primary vent tube during hard accelerations. They are featured on some carbs but are added easily to those without them.

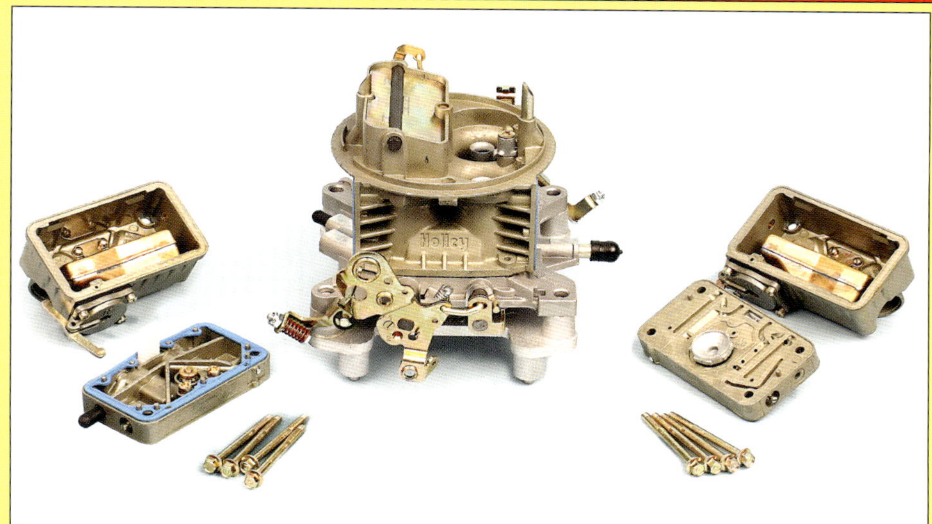

The primary and secondary fuel bowls and metering blocks have been removed on this typical 4150-series carb. This is a 750-cfm Double Pumper (PN 4779).

TECH TIP

4150/4160 Interchangeability

Parts that are interchangeable between the 4150 and 4160 series include jets, power valves, floats, needle/seat assemblies, vacuum diaphragm springs (for those equipped with vacuum secondaries), fuel bowl vent components, some accelerator pump hardware, primary bowl gaskets, and, of course, carb-to-manifold gaskets. ■

Double Pumper

If a 4150 carb has mechanical secondary operation, it has both a primary and secondary accelerator pump, and is referred to as a Double Pumper carb. Depending on the model, 4-barrel carburetors have either vacuum-operated secondaries or mechanical secondaries.

All primary sides include an accelerator pump, which is activated by the throttle linkage. The pump serves to inject an added boost of fuel to eliminate bog or lag when the throttles are opened, which is especially critical at lower engine speeds.

With a vacuum secondary design, the secondaries begin to open according to vacuum developed in one of the primary bore venturis, as engine load increases. During quick accelerator operation under load, a slight delay may be encountered while the secondaries begin to open. You can tune the vacuum sec-

The 650-cfm 4150 (PN 76650BL) is one of Holley's Double Pumper Ultra series, featuring billet aluminum metering blocks and a billet aluminum baseplate. The Ultra series is available with various color-coded blocks and baseplate in blue, red, and Hard Core Black.

ondary by changing to a lighter or heavier spring inside the vacuum diaphragm housing.

Double Pumper carburetors are equipped with mechanical secondaries; a second, separate accelerator pump circuit is located on the secondary side. The secondaries receive an immediate shot of initial fuel from the secondary's accelerator pump as the throttle linkage progresses.

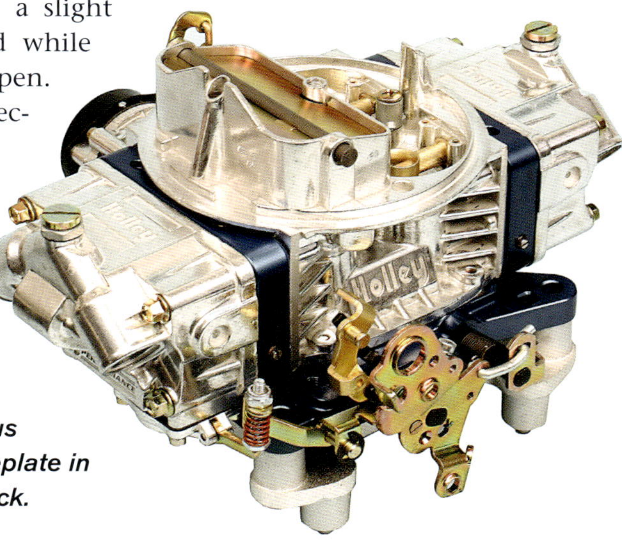

The Ultra Series 4150 carbs are offered in either mechanical- or electric-choke versions. The Ultra HP series eliminates the choke provision.

Double Pumper carbs are best suited to lighter-weight vehicles, those equipped with a manual transmission, and those with high–stall-speed automatic transmissions in the 3,000-rpm range.

Avenger

Introduced In 2009, the HP-series Street Avenger includes cast aluminum bodies with billet metering blocks and baseplates that reduce overall weight by an astounding 5 pounds compared to previous 4150 carbs. This line of carbs is also offered in both diecast and aluminum body construction. The Ultra Avenger series offers some appearance enhancements such as metering blocks and bodies in Hard Core Black, Hard Core Gray, blue, and red. The lineup is quite extensive.

In 2011, Holley released a new aluminum Ultra HP 4150-series carb with 30 new and improved features, including larger 1.6-inch venturis. The Avenger series is offered in ratings from 470 to 870 cfm, in both street and truck/off-road configurations. Although the majority of Avenger carbs have electric choke, six versions within the entire lineup are offered with manual choke. All Avengers have vacuum secondaries.

Street Avenger carbs were designed and calibrated specifically for optimal street performance, with features including a secondary metering block, four vacuum ports (for positive crankcase ventilation, power brake, spark, and vacuum accessories). The Avenger series is not designed for immediate bolt-on use with Chrysler automatic overdrive transmissions. If used with a GM overdrive TH700R4 or TH200R4 transmission, you need Holley's transmission kickdown cable bracket (PN 20-95) and geometry corrector (PN 20-121). Depending on your throttle bracket, you also need a throttle stud (PN 20-2, 20-38, or 20-40). The Street Avenger includes a Ford automatic transmission kickdown, but not for use with overdrive transmissions.

The Ultra Street Avenger series has aluminum construction and anodized billet aluminum metering blocks and baseplate (available in various colors). The Ultra is lighter (for those concerned with minimizing weight) and comes in several anodized colors. Personally, I love the Ultra series; in addition to the benefit of reduced weight, they look way cool.

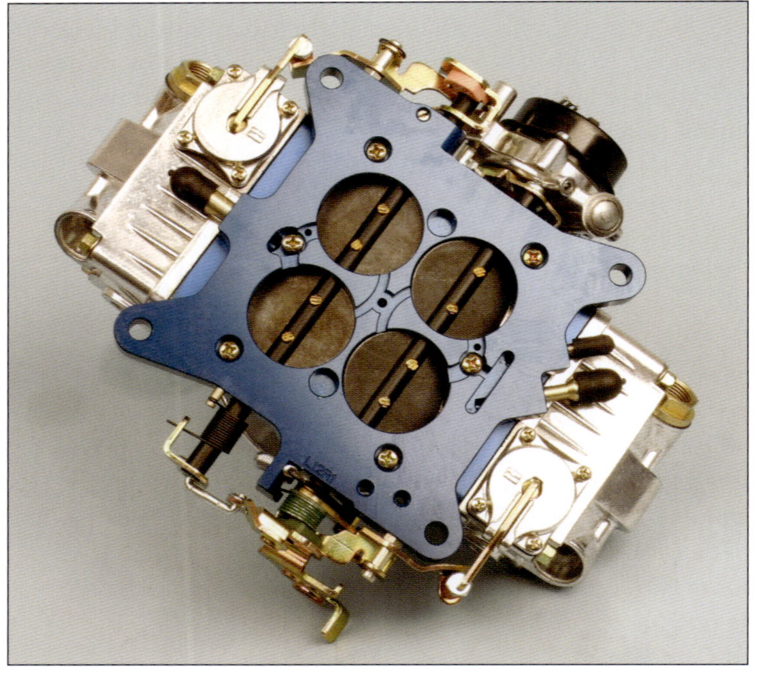

In addition to reduced weight and the obvious visual appeal, the anodizing protects the aluminum from long-term corrosion.

The Ultra Series 4150 carbs has convenient glass fuel bowl sights, making it much easier to check fuel bowl level during adjustments without the need to remove a solid brass plug.

In recent years, black has become the go-to color for those who wish to break out of the chrome-and-color mold and create a pro, understated, somewhat monochromatic appearance. Holley's Hard Core Black finish has become very popular. This is a 750-cfm Ultra Double Pumper (PN 76750HB).

In the Ultra series, the two letters at the end of the part number indicate color (HB is Hard Core Black, RD is red, BL is blue, etc.).

Holley Ultra 4150 street/strip Double Pumper carbs share the same features, including aluminum construction for reduced weight, billet aluminum metering blocks and baseplate, mechanical secondary operation, choice of manual or electric choke, four-corner idle adjustment, clear sight windows on both bowls, are optimized for out-of-the-box street/strip calibration, and are available in 650 to 750 cfm.

The Ultra HP series Double Pumper carbs have no choke and are designed for racing applications for which a choke is not required. This is a tumble-polished aluminum body and billet aluminum metering blocks and baseplate anodized in red (PN 80801RD).

All Ultra and Ultra HP carbs have double-pumper operation with an accelerator pump for each fuel bowl.

The Street HP carbs have a smooth, contoured venturi area for superior airflow. The boosters, the dichromate finished components directly over each venturi, are the down-leg style with a single fuel hole inside each barrel opening.

The Street HP series is essentially a 4150 race carb, using features taken from race HP and Double Pumper carbs, and then "tamed and tuned" for street and light-competition applications. The Street HP series is offered in yellow chromate, black, or tumble-polished finish. Sizes range from 600 to 950 cfm. The metering system includes a timed-spark port for a vacuum-advance distributor and it is calibrated for superior throttle response for street/strip use. The Street HP baseplate offers several vacuum ports for power brakes, positive crankcase ventilation (PCV), etc.

Although various finishes/colors are offered in Holley's current lineup, Hard Core Black (often referred to as Hard Core Gray) has become increasingly popular for those who wish to deviate from bright colors. This Ultra HP carb is mounted to an intake manifold that has been smoothed and hydro-dipped for a carbon-fiber appearance.

Note the absence of a secondary accelerator pump on the rear bowl on this Street Avenger. Remember, just because a carb has a dual feed fuel fitting for each bowl, it's not a Double Pumper unless it has an accelerator pump on each bowl.

Street Avenger carbs are specially designed and calibrated for optimum street performance; they are available in sizes ranging from 570 to 870 cfm. All have vacuum secondary operation and are available with either manual or electric chokes. The aluminum construction reduces weight.

Avenger carbs have an accelerator pump only at the primary bowl location.

Note the dual fuel-feed fittings on the primary and secondary bowls. However, the Avenger is not a double-pumper carb, so don't assume that a carb is a Double Pumper just because it has center-hung fuel bowls and dual feed.

Avengers are 4150-series carbs, so the secondary side has metering blocks with serviceable jets.

A quick-change vacuum housing cover allows changing the vacuum spring without the need to disassemble the entire housing unit. The cover is secured with two screws. Simply remove the two screws and then remove the cover and spring. Some carbs, such as the Avenger, have this as standard equipment.

The quick-change feature can be retrofitted to other carbs with a secondary vacuum kit. The housing cover and spring simply pull out. An O-ring seats onto the top of the housing. If it pops loose, be sure to reinstall it into its groove. The vacuum diaphragm spring sticks onto the housing top and is easily changed to a lighter or heavier spring to tune the secondary opening rate.

For off-road truck applications where severe vehicle operating angles are routinely encountered, Holley's Truck Avenger carb has a main vent tube that connects both primary- and secondary-side main vents. This prevents excess fuel spillover from sloshing out of the vent tubes. This is a 670-cfm Truck Avenger (PN 90670). (Photo Courtesy Holley Performance Products)

The Avenger Truck series is a unique design, intended specifically for truck and off-road applications. It has a single fuel inlet, square fuel bowls with side-hung floats, and an exclusive metering block and vent tube designed to prevent fuel spill-over at extreme angles. Flood-free operation accommodates climbing angles up to 40 degrees as well as side-hill maneuvers and nose-down descents up to 30 degrees.

The Dominator is often referred to as the "elephant" carb because of its large footprint and its large range of CFM ratings. The Dominator 4500 series ranges from 750 cfm all the way to 1,475 cfm.

Dominator

The "big boy" of Holley carburetors is the venerable Holley Dominator series, known as the 4500 series; it is often referred to as the "elephant" carb. The Dominator was designed as a large-airflow racing carburetor and debuted in 1969 (although some sources say 1968), initially to address the needs of NASCAR competition. Shortly thereafter, it was offered to the public. Today, the 4500 is extremely popular in drag racing applications. Versions are offered in 1,050, 1,150, and 1,250 cfm, as well as a "street" version rated at 750 cfm that was introduced basically to satisfy hobbyists who hungered for the Dominator look.

The Dominator flange is unique to the 4500 series. You must use a manifold designed for use with a Dominator or install an adapter mounting plate to accept a

Dominator. A Dominator cannot bolt to an intake manifold that is not designed for its larger bolt pattern.

Float bowls and metering blocks are similar to those found in the 4150 series, but the 4500 features an appreciably larger main body with the unique secondary linkage located between the front and rear throttle bores, underneath the main body. Throttle bores have a hefty 2.00-inch diameter, compared to 4150/4160

The Dominator series was developed specifically for top-end horsepower and torque. The classic Dominator has an aluminum main body with contoured venturi inlets for balanced airflow, a high-flow metering system, and annular boosters. The fuel bowls allow inlet plumbing on either side. The classic Dominator is offered in two- or three-circuit configurations and choices of 1 x 4 or 2 x 4 setups.

The new Gen 3 Dominator line represents an evolution in Dominator design with a 5/16-inch–taller main body that allows larger-radius air entries and fully machined venturi surfaces. It also includes 20 percent larger fuel bowls to eliminate starvation, 12-hole billet booster inserts, adjustable external linkage, and provisions for mounting a throttle position sensor. Annular boosters on the Dominator have a series of fuel orifices spaced around the inside diameter of each booster for superior fuel atomization. Boosters help to pull fuel from the bowls as air passes through them.

The Dominator 4500 has a non-staggered, square bolt-hole pattern. Other 4-barrel models have a staggered pattern with the bolt holes spaced farther apart from front to rear when compared to left-to-right bolt holes.

Throttle bore diameters on 4150- and 4160-series carbs measure 1.563 to 1.750 inches, depending on specific part number.

The throttle bore diameter on Dominator (4500-series) carbs measures 2.00 inches.

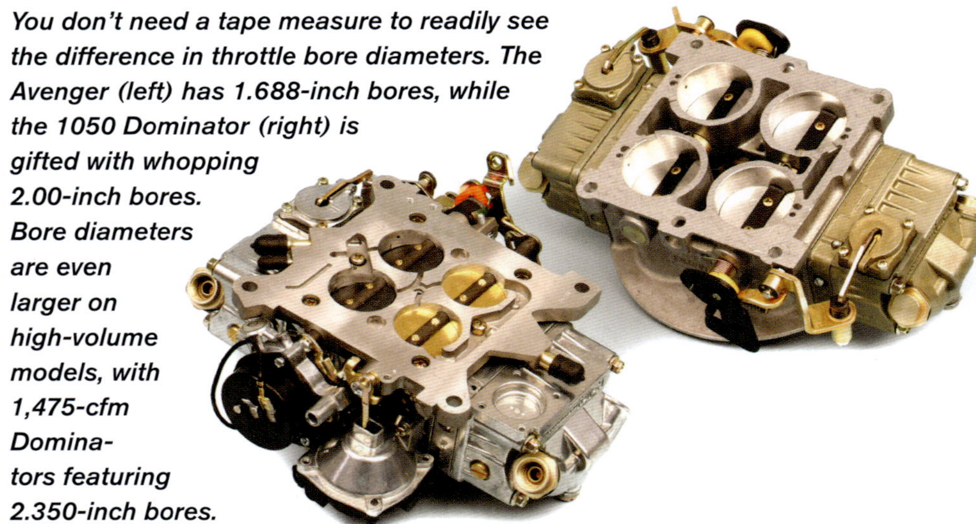

You don't need a tape measure to readily see the difference in throttle bore diameters. The Avenger (left) has 1.688-inch bores, while the 1050 Dominator (right) is gifted with whopping 2.00-inch bores. Bore diameters are even larger on high-volume models, with 1,475-cfm Dominators featuring 2.350-inch bores.

diameters that generally range from 1.563 to 1.688 inches.

Baseplate Bolt Pattern

All Holley 4-barrel carburetors (except for the 4500 series) have a baseplate bolt pattern with the baseplate-to-intake manifold bolt holes stagger-spaced 5¾₆ (left to right) x 5⅝ inches (front to rear). As previously mentioned, the 4500-series Dominator baseplate bolt pattern is unique to that series and includes a square bolt pattern measuring 5.375 x 5.375 inches. The Dominator requires the use of an intake manifold designed to accept a Dominator. However, adapters are available that allow you to mount a 4150/4160 to a Dominator manifold and vice versa. Depending on the specific intake manifold, installing a Dominator onto a manifold designed for the

Holley Carburetor Models and Types

Holley carburetors are available in different "series" designs. Within each family are variations in CFM, choke design, secondary design (on 4-barrel carbs), etc.

The 4160-, 4150-, and 4500-series carburetors are, by far, the most commonly used. Depending on the engine size and application, carbs in these three series can be applied to street vehicles. However, generally speaking, the 4160 is intended for the street, the 4150 is for serious street and racing, and the 4500 is designed primarily for racing.

For a complete listing of specific Holley carburetor models and the vehicles on which they were installed, please refer to the Appendix.

Series 4160

The 4160 includes a single fuel inlet with a transfer tube that feeds the vacuum-operated secondaries and a secondary metering plate rather than a secondary metering block. The metering plate is non-adjustable and has no replaceable jets. A power valve and an accelerator pump are located on the primary side only. The 4160 carburetors are the less-expensive versions of the 4150s but can be upgraded easily to 4150 status by installing a secondary bowl and metering block kit.

Series 4150

The 4150 carburetors are similar to 4160 carbs; their primary difference lies with secondary operation. The 4150 is enhanced with features such as dual fuel inlets, vacuum or mechanical secondary operation, metering blocks (with replaceable jets) on both the primary and secondary sides, and a power valve on both the primary and secondary sides. The 4150 series offers a much wider range of CFM sizes compared to the more "basic" 4160.

Series 4500

Famously known as the "Dominator" series, the 4500 is the big boy on the block. Notable differences between the 4150 series and the 4500 series are the larger 2.00-inch bores, a secondary throttle linkage that runs internally, throttle-shaft openings sealed to prevent contamination, and the elimination of an air horn/air choke to control air entry. The Dominator series has a larger bolt pattern for intake manifold mounting. Although adapters are available to mount a 4500 carb onto a 4150 intake manifold, the proper choice is a manifold that is designed to operate with a 4500-series carburetor. Dominators are good choices for engines producing more than 625 hp.

Series 4165

This series is known as a Double Pumper because it features mechanical secondaries, an accelerator pump at both bowls, and a metering block with jets at both primary and secondary sides. It has a spread bore pattern.

Series 4175

These carburetors have a spread-bore base and are designed as a direct replacement for the GM Quadrajet. Series 4175 carbs have electric chokes and a single fuel inlet.

Series 4010

These carbs have a square bore with either mechanical or vacuum secondaries.

Series 4011

This series is essentially the same as a 4010, but is designed for spread-bore intake manifolds. It is offered with either mechanical or vacuum secondary operation.

Series 4224

Built as a 660-cfm drag racing carb, this carb has a center-squirter design that squirts fuel into all four barrels at the same time, aided by a hefty 50-cc accelerator pump. ∎

The 4150 and 4160 bolt spacing, from front to rear measures 5⅝ inches.

Dominator bolt spacing, from left to right, is specified at 5⅝ inches.

Dominator bolt spacing, from front to rear, is specified at 5⅝ inches.

4150/4160 carbs may require enlarging the manifold's plenum ports.

Air-Cleaner Base

The 4150/4160 series requires a 5⅛-inch air cleaner/gasket; the larger Dominator requires a 7⁵⁄₁₆-inch air cleaner and gasket. These are the standardized published sizes that appear in all catalogs; they provide a comfortable drop-on fit. Actually, the 4150/4160 carb air cleaner inside-diameter gasket surface measures approximately 5.001 inches; the inside-diameter gasket surface of the Dominator measures approximately 7.251 inches.

Air Cleaner Threads

Most Holley carbs, including the 4150/4160 series, have a 1/4-20 threaded center hole to accept an air cleaner stud. The Dominator (4500) series has a larger 5/16-18 threaded hole. Inexpensive male/female adapters are readily available from any performance retailer to allow the use of either stud size. Adapters are available with 1/4-20 male thread and 5/16-18 female thread as well as 5/16-18 male thread and 1/4-20 female thread.

HOLLEY CARBURETOR OPERATION

Holley 4-barrel carburetors are dual-stage, downdraft units. A dual-stage carb has a primary side that supplies the air/fuel mixture throughout the entire stage of engine operation, and a secondary side that operates only when the engine demands a greater quantity of air and fuel.

The basic components include a main body, baseplate (or "throttle body"), primary and secondary fuel bowls fitted with adjustable floats, accelerator pump and accelerator pump nozzle (or "squirter") on the primary bowl, primary and booster venturi with primary discharge nozzle, throttle plates, idle fuel passage, and a primary metering block with screw-in jets on the primary side. The 4150 series has a secondary metering block; the 4160 has a secondary metering plate. The 4150s with mechanical secondary operation have an accelerator pump and squirter on the secondary bowl.

The venturi (or "barrel") passage has a double-taper design: The larger area at the top narrows and then enlarges above the throttle plate. This shape creates higher velocity and lower pressure at the narrow section; the vacuum effect draws in fuel and air.

The fuel mixture is pre-atomized, mixing with air from the small air-bleed orifices in the top of the main body as it passes through the discharge nozzle in the venturi booster. It is emulsified further as it passes through the high-speed column of air, resulting in the air/fuel mix to change the fuel stream into tiny droplets in a "mist" configuration. The air bleeds are sized to work in conjunction with the vacuum effect in the venturi (called the venturi effect), which increases as the throttle is opened and is greatest at wide-open throttle (WOT). The venturi and booster venturi work

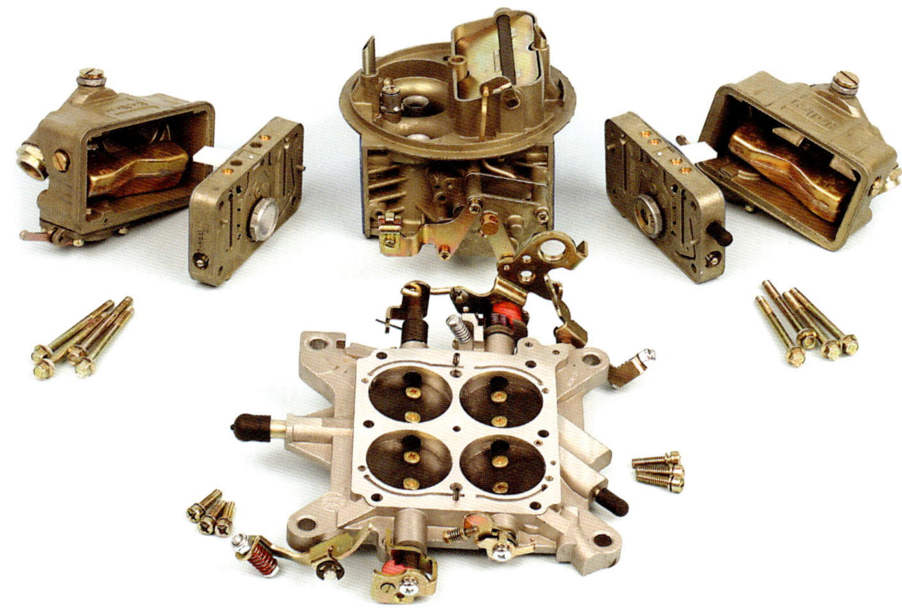

This Holley 4-barrel carb 4150 (PN 4779) is disassembled into its basic components. This carb has mechanical secondaries and a manual choke. At the bottom center is the baseplate, also called the throttle body. At the top center is the main body. The primary fuel bowl and metering block are at the right and the secondary fuel bowl and metering block are at the left.

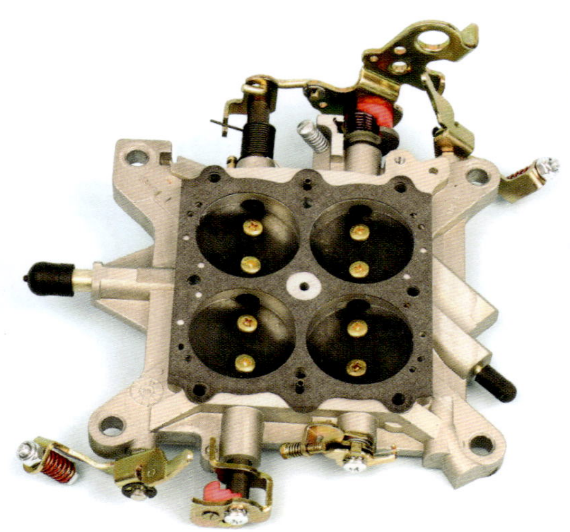

This is a baseplate from a 4150 Double Pumper. The baseplate, also known as the throttle body, is the foundation for a carb and incorporates the throttle shafts and throttle plates.

Venturi boosters come in different designs. These are "down-leg" boosters; note the angle of the passage body. The inside of the booster air passage hole is a small fuel orifice.

The fuel hole in this straight-leg booster is located on the underside of the small passage pipe that runs through the center of the booster air passage.

together; the booster venturi aids in generating additional pressure drop at reduced air speed and increasing the fuel flow.

Booster Venturi

Several booster designs have been used in Holley carburetors. They provide more or less restriction; signal strength increases as restriction increases, and this affects the pressure drop. In addition, fuel distribution and degree of atomization vary according to the signal.

A "standard" booster used today has a straight leg and a single small orifice passage for fuel entry into the booster's cavity. A "down-leg" booster includes an angled leg, which provides less restriction.

Annular boosters, found in some 4150 series and all 4500 Dominator carbs, have a series of small orifices around the inside of the booster. They create a stronger signal that is especially useful with engines that have long-duration camshafts and low manifold vacuum at low engine speeds.

This 4500 Dominator is equipped with annular boosters. The inside of the booster's main air passage has a series of fuel orifices that runs around the inside diameter of the air passage.

Holley carbs come in many designs. The 4160s have a fuel inlet at the primary bowl; a transfer tube then supplies fuel to the secondary bowl. The 4150s include dual fuel inlets, with an inlet at each bowl. Fuel travels through the needle and seat assembly and the bowl fills.

Regardless of series, as the fuel level inside the bowl rises, the float rises. Pressure from the float causes the needle to seat, cutting off the fuel flow. As the engine uses the fuel, the fuel level drops, lowering the float. Pressure from the dropping float unseats the needle, allowing fuel to flow into the bowl once again. The float level must be adjusted correctly

Here, you can see some of the small fuel orifices inside the annular booster.

so that the basic fuel metering system operates as designed.

On carburetors equipped with center-hung floats, float adjustment is handled externally via needle height adjustment. To make an adjustment, loosen the slotted plug (which is

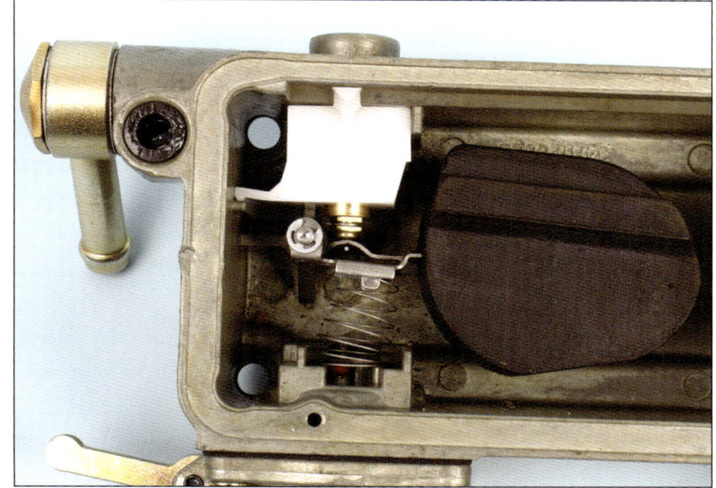

This close-up of a side-hung float clearly shows how the needle and seat assembly contact the float lever arm.

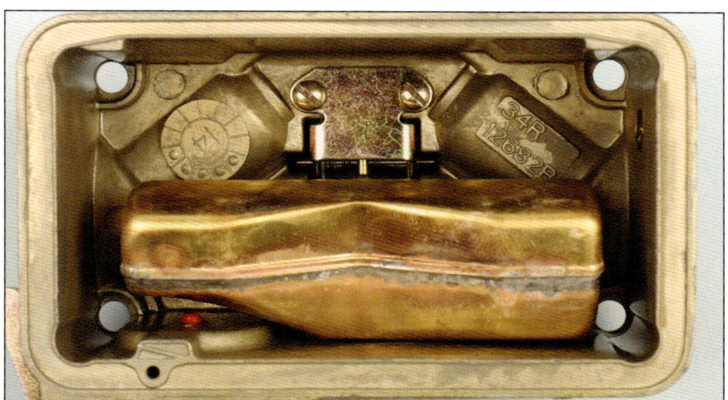

Here is a center-hung float in a 4150 fuel bowl. The needle and seat assembly contact the top of the float, which hinges from a center-mounted location. This style provides more consistent float and fuel level in the bowl during both acceleration and lateral maneuvers.

The top of the main body has air-bleed orifices on both the primary and the secondary sides of a 4150- or 4500-series carb. The outboard brass holes are for the primary idle air bleeds.

Although the air bleeds on most carbs are press-fit, the 4500-series Dominators have slotted and replaceable air bleeds for fine-tuning. Using larger diameter orifices creates a leaner air/fuel mixture.

The locations of the air bleeds on the Dominator are the same as on other models. The inboard holes are the high-speed bleeds, and the outboard holes are for primary idle air bleed.

Center-hung float bowls have a needle and seat assembly located at the top center of the bowl. The fuel inlet is located on the passenger side of the bowl.

The hex nut engages to the needle and seat assembly via a rectangular engagement. A 5/8-inch wrench is required to drive the hex nut. Turning the hex nut counterclockwise raises the needle and turning the nut clockwise lowers the needle. Whenever the hex nut is removed, always inspect the gasket on the underside of the nut and replace if it is damaged. A slotted screw serves only to lock the nut in place.

Removing or installing the needle and seat assembly must be done with care to avoid damaging the assembly's O-ring.

the lock), turn the hex to make the adjustment, and then tighten the lock. A sight port on the side of the fuel bowl allows you to monitor fuel height while the engine idles.

Some carbs have a brass screw-in plug that must be removed. Fuel level is adjusted until the fuel is at the bottom of the sight hole. Other models have a glass sight window, which allows you to observe the fuel level without having to remove a plug. On 4160s with side-hung floats, the fuel bowl must be removed and turned upside-down; adjust the float so that the top of the float surface is parallel to the roof of the bowl.

This is a front view of a metering block equipped with main jets. The slot on each jet allows servicing with a flat-blade screwdriver, although a dedicated Holley jet driver is preferred because it avoids accidental slot deformation. See Chapter 4 for information about specialty carburetor tools.

Main jets are available in a wide range of orifice sizes. All jets share the same exterior shape, size, and thread. A selection of jet sizes is shown here, threaded into a billet aluminum jet card from AED. A jet organizer such as this allows a neat and orderly method of storing extra jets if you plan extensive tuning.

Each fuel bowl is vented internally to the air horn by a vertical vent tube in the main body, which allows the release of excess fuel vapor.

Jets

Screw-in metering jets are exactly what the name suggests. They meter the amount of fuel that flows from the fuel bowl into the main fuel circuit; larger metering holes allow more fuel to flow and smaller jets

Main jets are machined from solid brass. The metering orifice sizes range from .040 (jet number 40) through .128 inch (jet number 100). Selection in jet sizes is mostly in .001-inch increments.

allow less fuel to flow. Fuel runs through the jets into the metering block and then upward along the channels to where it meets with air from the air bleeds. At this point, the emulsification process begins and the fuel mixture enters the main body. The pressure drop in the booster venturi pulls air from the air bleeds and pulls fuel from the jets.

Idle Circuit

The idle circuit supplies the air/fuel mixture for engine operation at

The jet number is stamped into the side of each jet. This jet is a number 71, which indicates that the metering orifice is .076 inch.

The main jets have been removed from the metering block, but the power valve is in place. The main jet thread's size is identical for all jets, at 1/4-32. Never add any type of thread-locking compound to jet threads. Jets are tightened snugly to 30 to 40 in-lbs.

idle and low engine speed. The purpose of a secondary idle circuit is to maintain a constant fuel level in the secondary fuel bowl. Both primary barrels use identical idle circuits and both secondary barrels use identical idle circuits. However, the primary and secondary idle circuits do differ from each other.

Primary Idle

Fuel flows from the primary fuel bowl through the main jets, into a small, horizontal idle-feed passage that leads to a vertical idle-well passage. It then flows past an idle feed restriction, through another horizontal passage, and is eventually mixed with incoming air from the idle air bleed. The fuel then flows down a vertical passage to the bottom of the main body, where it splits in two directions; one path goes to the idle discharge passage and the other goes to the idle transfer passage and constant-feed port. The mixture that flows to the idle-discharge passage flows past the tip of the idle mixture adjustment needle screw, then through the main body and to the throttle body, where the mixture is discharged into the throttle bore, below the closed throttle plate.

The mixture flows unrestricted in the passages leading to the idle-transfer passage and constant-feed port, through which it exits. When the throttle plate is closed, no fuel is discharged through the idle transfer slot. The transfer slot acts as an air bleed directly above the idle constant-feed port and serves to further lean out the air/fuel mixture.

As the throttle plate opens and engine speed increases, the idle-transfer slot is exposed to intake manifold vacuum and fuel is discharged from the transfer slot. As

The vertical idle air bleed in the metering block intersects the short horizontal passage that locates the idle-mixture screw.

the throttle plate continues to open, engine speed and airflow through the carburetor increases; it is increased even further by the venturi effect. As airflow increases, the main metering system begins to operate and the idle system begins to taper off; this provides a smooth transition from idle to engine operating speeds.

Secondary Idle

All Holley 4-barrel carburetors have an idle circuit built into both the secondary and the primary sides. Some secondary idle circuits are pre-

determined by design and are not adjustable, but some 4150 carbs with mechanical secondaries have the same idle adjustments that are found in the primary side. Some (but not all) carb models in the 4150 series have idle-mixture screws in the secondary, as well as in the primary, metering blocks. This feature is provided on an individual basis according to the part number.

An idle circuit in the secondary side controls fuel flow through the secondary bowl; it helps maintain proper fuel level, even if the secondary throttle plates are not open. To set up a carb with both primary and secondary idle-mixture screws, gently screw them in until they seat. Then, back them out one full turn. When the engine is running, adjust the idle speed. Depending on the design of the intake manifold, it is easy to tune for a variety of specific applications with this four-corner mixture adjustment setup.

Main Metering

When the throttle is partially open, the main metering system

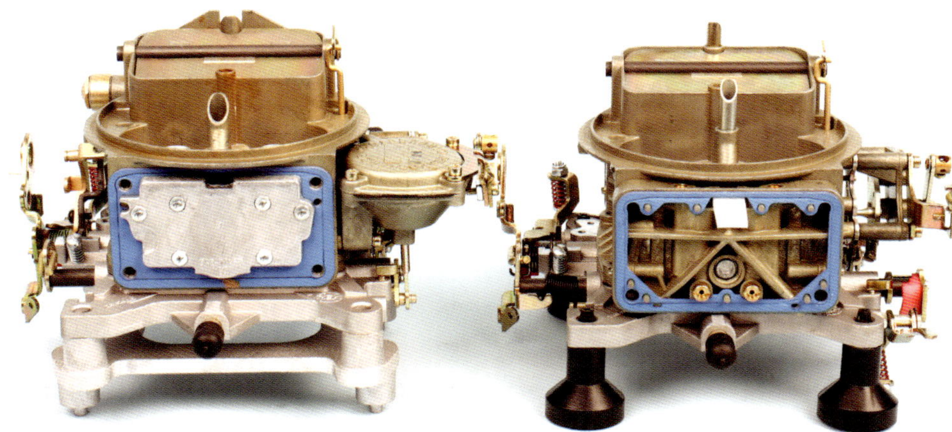

This comparison of a 4160 carb (left) and a 4150 carb (right) clearly shows the secondary metering plate on the 4160 and the secondary metering block on the 4150.

on the primary side meters the fuel flow from the fuel bowl through the main jets and into the main fuel well. Fuel flows past the main-well air bleeds, where it mixes with air. This air/fuel mixture then exits through a horizontal passage to the discharge nozzle in the booster venturi, where it mixes with incoming air. Because this air/fuel mixture is lighter than liquid fuel, it responds faster to changes in venturi vacuum and vaporizes more easily when discharged into the airstream. Throttle plate position regulates the amount of the air/fuel mixture entering the intake manifold, and as a result, it regulates engine speed.

On the secondary side, the main metering system operates in a similar manner. On 4160 carbs, fixed (or predetermined) restrictions are machined into the metering plate. Idle fuel wells branch off each main well. Fuel for the idle and idle transfer system enters the main well through the plate restrictions and then it travels through the idle well and the idle restriction. Finally, it is mixed with air entering from the secondary air bleeds.

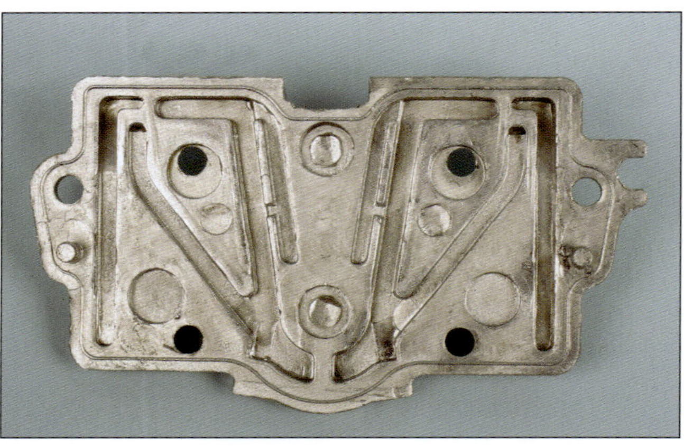

With a metering plate, the fuel enters the metering holes at the bottom of the plate. At idle, fuel travels up from the metering holes along the angled "dogleg" passages, through the idle feed passages at the top of the doglegs, and then down through the vertical passages on each side of the plate to the idle transfer slot in the main body. When the fuel moves through the main circuit, it travels from the main metering holes upward through the inboard (slightly angled) passages where the fuel is emulsified by air from the main body. From there, it moves through the main body to the discharge nozzles and into the venturis.

Note the main metering holes at the bottom of this secondary metering plate.

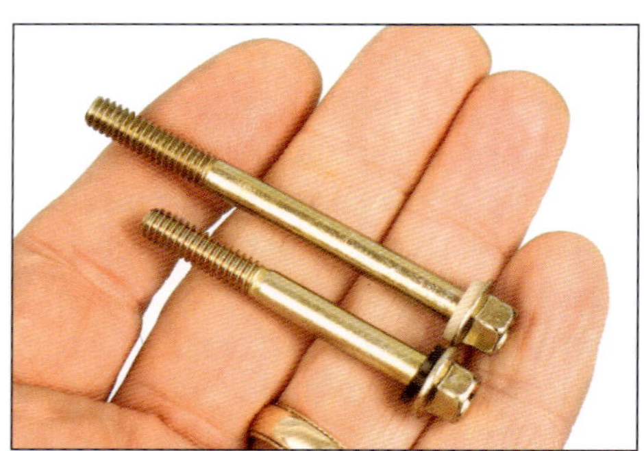

All fuel-bowl screws are thread size 12-24 and are available in two lengths. The 4150- and 4500-series primary and secondary fuel bowls, as well as the 4160 primary fuel bowl, use 2½-inch screws (top). The secondary fuel bowl on the 4160 carb (with the thinner metering plate) uses 1⅞-inch screws (bottom).

Six 1/2-inch 8-32 screws secure the secondary metering plate to the main body on a 4160 carb. These screws are a clutch-head drive.

Fuel-bowl screws seal to the fuel bowls with fiber or nylon washers. Inspect these sealing washers carefully every time you remove or install bowl screws. If you plan to service the carburetor on a regular basis for tuning or general service, it's a good idea to have new bowl screw washers on hand.

The underside of the baseplate has crossover channels, as well as small holes, called curb-idle discharge ports, which are in the throttle body ports under the throttle plates. Engine vacuum pulls a small amount of fuel from the discharge ports as part of the idle circuit.

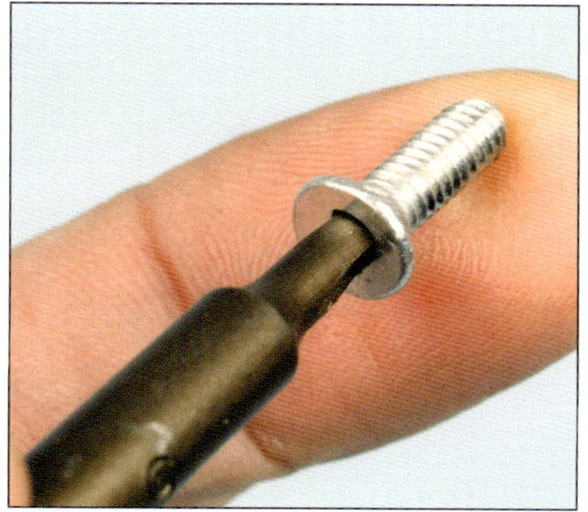

The clutch-head screw drive looks like an hourglass. Although you can use a flat-blade screwdriver, you should, more correctly, use a clutch-head driver. A 5/32-inch clutch-head driver is available as a bit to attach to a drive handle or ratchet or as a dedicated driver with a full-size grip. The clutch-head driver engages the screw's drive and provides secure removal or installation without the chance of damaging the screw's drive. This full-size clutch-head driver is from NAPA (PN M-135).

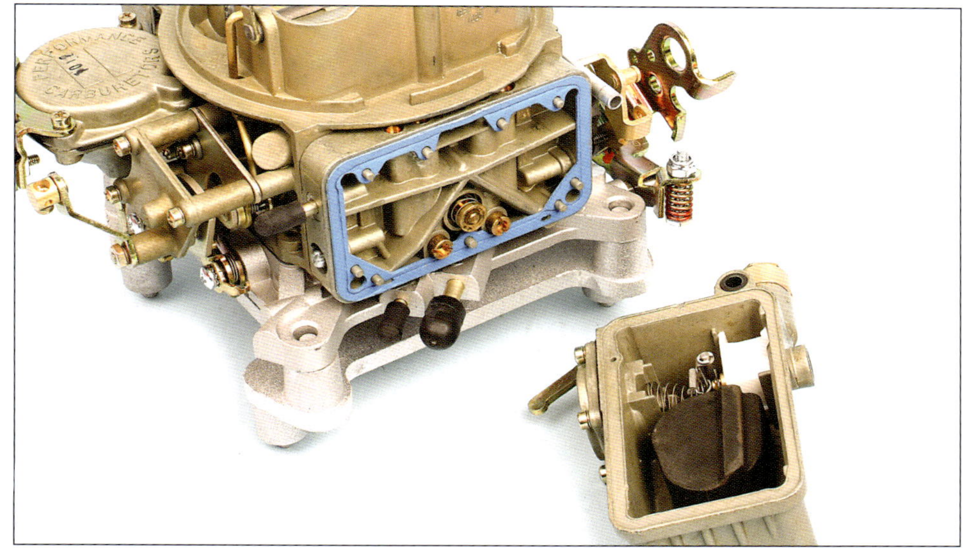

The primary bowl has been removed from this 4160 carb. Its primary metering block is similar to that found on a 4150 model. Note the side-hung float used in 4160 models.

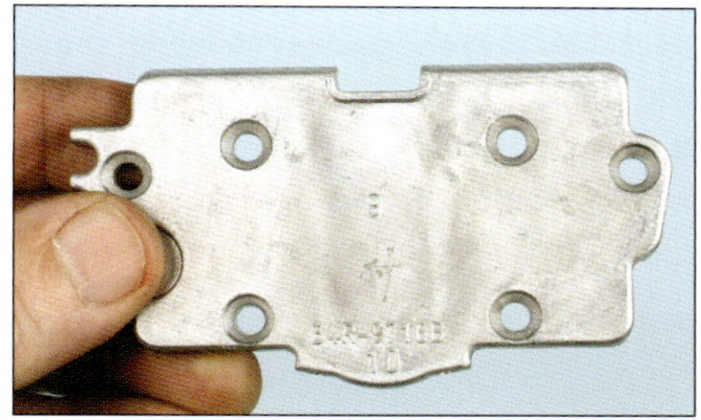

All Holley secondary metering plates are stamped with a one- or two-digit identification number in the center area. Production numbers, which can be ignored, are stamped at the center bottom. The "9" in the center identifies this plate as PN 134-9; its main metering holes are .067 inch and its idle holes are .031 inch. Metering plates are available in a wide range of primary and idle circuit sizes.

Shown here are the main bodies of a 4160 series (left) and a 4150 series (right).

The 4160 (left) uses a thin metering plate between the secondary fuel bowl and main body; the 4150 (right) uses a metering block. This side-by-side-comparison clearly shows why a 4160 requires shorter screws for its secondary fuel bowl.

Power Valve

During hard acceleration, the increase in airflow results in leaning the air/fuel mixture. A power valve is located in the primary metering block to provide the additional fuel to counter this and to enrich the mixture. The power valve has an internal diaphragm and is operated by vacuum supplied through passages in the throttle body baseplate and the main body. At idle or under load, there should be enough vacuum to keep the power valve closed.

As manifold vacuum drops under high-speed conditions, the spring in the power valve forces the diaphragm to open, allowing extra fuel to flow through the power valve. The fuel then travels through the restrictions in the metering block and into the main well, joining the main fuel flow, which richens the mixture. After engine speed is reduced, the manifold vacuum rises and causes the power-valve diaphragm to overcome its spring pressure and close.

On the left is a single-stage power valve with small orifices. On the right is a single-stage power valve with a large window passage. The type with small holes is primarily designed to improve part-throttle economy on vehicles with heavy loads and isn't the best choice for a performance application.

The gasket style differs as well. The correct gasket must be used depending on style of power valve. If the gasket is incorrectly matched to the power valve, it leaks. The gasket with a concentric inside diameter is used with a power valve that has the larger window fuel opening. The gasket with three small tangs on the inside diameter seats properly onto the power valve with the series of small fuel orifices.

The power valve remains closed during idle and normal cruising. The power valve opens when intake manifold vacuum drops to a certain level and load increases to supply additional fuel.

Here the power valve has been installed in the metering block. A 1-inch wrench is required for removal or installation. The flats on the valve are very shallow, so make sure that the wrench is seated fully. A superior choice is a specialty power-valve wrench (see Chapter 4 for details). The opening vacuum of power valves is marked in inches. This valve is marked 6.5, meaning that it opens when manifold vacuum drops to 6.5 inches.

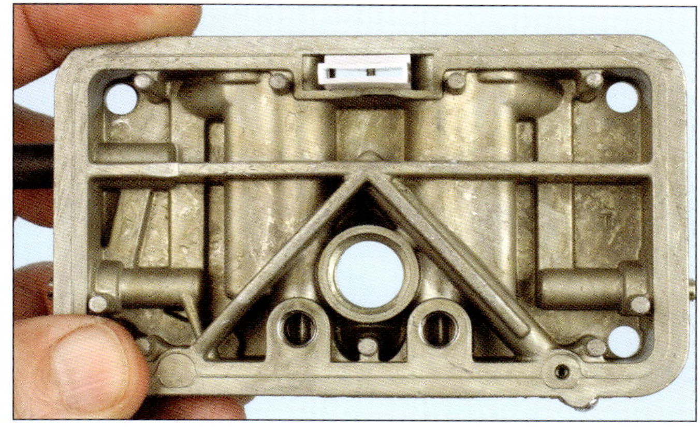

The jets and power valve have been removed from this metering block. The thread size of the power-valve hole is 1/2 x 28.

Whenever replacing a power valve, it's a good idea to install a new gasket. A power valve should be tightened to the metering block at 40 to 50 in-lbs. Avoid overtightening.

Venting

The fuel bowl is vented internally to the air horn by a vent tube in the carburetor body, which allows the escape of excess fuel vapors. With the primary fuel bowl removed, you can see a small rectangular passage at the top center face of the metering block. This opening aligns with a small rectangular passage in the main body that enters the vertical vent tube. If the float is set too high, or if the needle-and-seat assembly are stuck (or if a small crack is in the metering block or main body), fuel can spurt out of the vent tubes and send excess fuel into the throttle bores, causing a bog.

Under hard acceleration or when driving in off-road conditions where extreme vehicle angles are encountered, fuel from the primary bowl can suddenly be crammed into the vent passage and out of the vent tube. A vent "whistle" is standard on some Holley carbs, or one can be added easily to an existing carb. This vent whistle extends into the fuel bowl area. If, under hard acceleration, a shock of fuel is slammed into the passage, the excess fuel is diverted through the vent whistle and back into the fuel bowl.

Another method of avoiding excess fuel spill through the main vent tubes is with a radiused U-shaped crossover vent tube that connects to both the primary and secondary main vent tubes. This tube has small holes to allow air venting, but it provides a longer travel path if excess fuel tries to slosh out of the bowl through the main vents. This type of crossover vent tube is popular among off-roaders and for some marine applications where severe operating angles are commonly encountered.

Vertical vent tubes are located at the front and rear of the main body. Be aware of clearance issues between the top of the vent tubes and the air cleaner lid. Opinions vary regarding an acceptable clearance, but generally speaking, you should have at least 3/8 inch between the top of the vent tubes and the air cleaner lid. This may be a concern when dealing with a drop-base-style air cleaner where the lid is closer to the air horn. Some vent tubes have an angled cut at the top to minimize the risk of blocking the vent tubes. Other carbs (usually those with no choke) include shorter vent tubes that are straight-cut so they're short enough to stay well away from the air cleaner lid.

In addition, the primary and secondary sides of the main body each have four small holes at the top that

A white fuel-resistant plastic fuel–bowl vent baffle extension, known as a "whistle," aids in preventing fuel spillover during hard acceleration, deceleration, and cornering.

The main vent tubes are clearly visible on this 4150 carb.

This is a close-up of the primary vent tube.

Here you can see that the tops of the secondary vent tubes are cut at an angle. This provides a larger cross-sectional area for the vent and ensures venting even if the air cleaner lid is very close to the vent tips.

The vent tubes on this 4500 Dominator are angle-cut. If vent tubes have a square-cut tip, the air cleaner lid must be far enough above the tips to avoid venting restriction. A general rule is to have a minimum of about 3/8-inch clearance.

To avoid excess fuel spillover from exiting the primary and secondary main vents, a hoop-style vent tube may be included. This provides a longer travel for any fuel that may be spilled out of the vents due to operating at extreme angles (such as in severe off-road conditions), containing the fuel and allowing it to drain back to the fuel bowls. A series of small top orifices serve to vent the main vent circuits to the atmosphere. Some Holley carbs, such as this 90670 Truck Avenger, include this feature as standard. (Photo Courtesy Holley)

serve as air bleeds. The two inboard holes bleed the main circuit; the two outboard holes are for the idle circuit. These air bleeds must remain clean and free of debris.

Accelerator Pump

An accelerator pump is a small lever-operated diaphragm pump located on the underside of the fuel bowl that is connected to the throttle linkage. All Holley 4-barrel carburetors have a primary-side accelerator pump located on the driver's side of the primary fuel bowl. The 4150 carbs with mechanical secondary operation, and all 4500 series carbs, have an additional accelerator pump

This orange "dot" indicates the rubber/silicone umbrella-style check valve. This style, in contrast to the older check ball approach, seals with very little pressure, and moves from the sealed position when fuel is drawn into the accelerator pump cavity. Unless it's leaking, don't disturb it.

The accelerator pump discharge nozzle/squirter is located at the center on this primary side of a 4150 with the choke plate held open.

This is the secondary accelerator pump discharge nozzle/squirter on a 4150 Double Pumper carb.

on the passenger side of the secondary fuel bowl, which is why they are called Double Pumpers.

The accelerator pump provides a quick, brief shot of fuel into the venturi as soon as the throttle is moved. Fuel is sent via the pump action through the accelerator nozzle (squirter) that is located immediately above the venturis. The squirter nozzle has twin orifices, with one aimed at each barrel on the primary and/or secondary barrels.

During the initial hit on the throttle, engine vacuum isn't high enough to draw fuel through the venturis quickly enough, which results in a momentary stumble. The quick shot of fuel from the accelerator pump compensates and provides a smooth transition from initial throttle to wider throttle opening.

The pump's operation can be tuned by changing the pump itself, the pump's squirter nozzle, and/or the cam that's mounted on the throttle lever. The cam determines the amount of "lift" that takes place at the accelerator pump arm, similar to an engine's effective valve lift (the ratio of the rocker arm increases the valve lift above the camshaft lobe lift). A selection of profile plastic cams is available to tune the degrees of lift at the accelerator pump arm.

Variables, including vehicle drive gearing, vehicle weight, camshaft timing, carburetor size, engine torque, etc., influence the length of time involved in the transition period. As engine load is introduced, both airflow and engine vacuum are low, and the need for the shot of fuel from the accelerator pump increases. As the lag time increases, a longer shot of fuel is needed to maintain smooth transition as the throttle is opened.

These are a discharge nozzle, screw, and washers. The machine screw has a 12-28 thread with a length of 3/4 inch. Note the tapered underside of the screw's head that countersinks into the top of the nozzle screw hole. A thin metal washer is located between the nozzle and main body, and between the screw head and nozzle. The accelerator pump nozzle has twin squirt outlets, one for each barrel on a primary-side (and for each secondary barrel on a Double Pumper) carb.

Accelerator pumps are mounted to the bottom of the fuel bowl with four 8-32 machine screws. Most carb models have a screw length of 3/8 inch; the 4500-series Dominator uses 9/16-inch-long screws.

The accelerator pump lever is activated by an extension of the throttle lever. Use the spring-loaded screw for fine adjustment at the pump lever contact point.

With the accelerator pump housing removed, you can see the pump diaphragm.

If you plan to reuse the diaphragm, the sealing gasket may be stuck to the fuel bowl, so be careful during removal to avoid tearing the gasket.

Remove the accelerator pump diaphragm. The convex side faces the pump housing.

The concave side of the diaphragm faces the bowl. This recessed side captures the diaphragm spring.

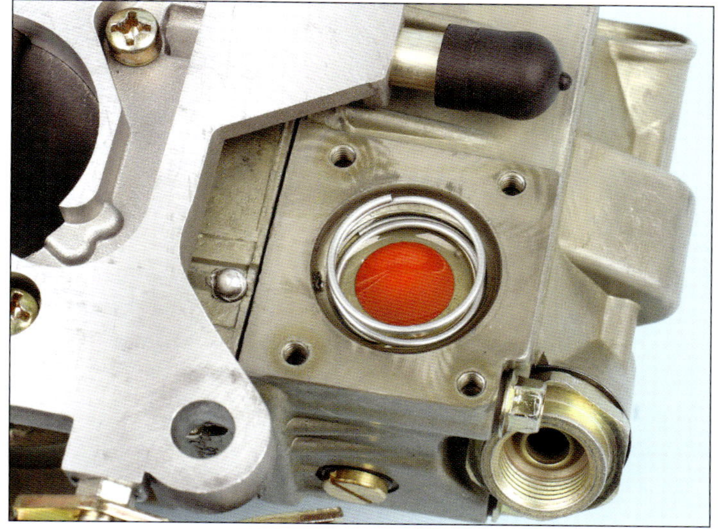

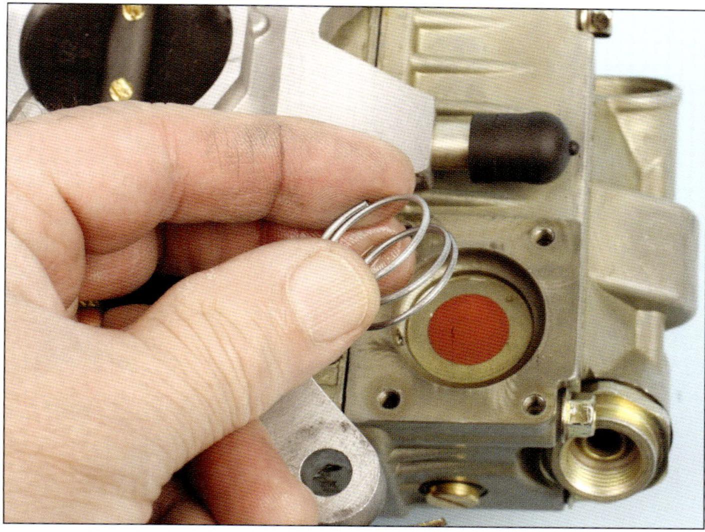

With the diaphragm removed, note the pump spring.

Remove the spring by hand.

Fuel Inlet

Depending on the specific carburetor, it has either a single or dual fuel inlet. The 4160 carbs typically have a single fuel inlet; fuel from the fuel pump enters only the primary fuel bowl and is then transferred to the rear bowl via an external fuel transfer tube. The 4150 carbs, whether they are Double Pumpers or not, have dual fuel inlets; a feed port is at each fuel bowl.

A single-feed inlet typically includes a banjo fitting that positions a barbed slip-on fuel hose leg at 90 degrees to the side of the fuel bowl. The banjo fitting has a sealing crush washer at each side of the round banjo end. One washer is between the banjo and the fuel bowl and another washer is between the banjo head and the bolt that secures the fitting to the fuel bowl. The banjo bolt runs through the round fitting head and has a fuel passage in the side of its shank. It also has an exit passage at the shank tip, allowing fuel to flow into the banjo fitting, turn 90 degrees, and then enter the fuel bowl.

Dual-feed carburetors that have a fuel inlet for each fuel bowl are found on carbs equipped with center-hung floats. A fitting with male threads engages the female-threaded hole in the fuel bowl; it also has a female-threaded entry hole that accommodates a line/fuel log fitting.

In a 4160 fuel bowl the female thread size that accepts the banjo fitting is 9/16-24. The fuel inlet fitting thread size in each center-hung fuel bowl is 7/8-20. Depending on how

Fuel inlet fittings on center-hung fuel bowls have a 7/8-20 thread at the fitting-to-bowl connection. The common hex size requires a 1-inch wrench. A thin metal washer seals the fitting to the bowl.

The standard inlet fitting has an inverted flare to accept a double-flared fuel line. Various adaptations are available to accommodate –AN fittings either to the inlet fitting or to take the place of the stock inlet fitting.

Some carbs, such as the Ultra Series 4150 carbs, have the same thread size inlet fittings, but with a smaller hex for easier wrench access. These fittings require a 3/4-inch wrench.

This 4160 carb includes a single inlet fitting at the primary bowl, utilizing a banjo-style fitting. The angle of the hose nipple can be adjusted by simply loosening the banjo bolt and retightening it in the position of choice. With the banjo fitting removed, you can see the in-line filter screen attached to the end of the banjo bolt shank.

The banjo bolt fuel inlet assembly consists of the banjo, banjo bolt, filter, and two crush washers.

The banjo has a tube inlet nipple (for fuel hose connection). The fuel hole enters the inside of the bolt passage.

The banjo bolt has a transfer hole in its hollow shank. When fuel exits the banjo, it enters the side of the banjo bolt and exits through the bolt shank, through the filter, and into the fuel bowl. If you wish to modify the plumbing, banjo assemblies are also available that accept –AN hose (if you prefer a threaded –AN hose connection rather than a slip-on hose). Or, if you wish to eliminate the 90-degree banjo approach, a variety of –AN fittings are also available that thread directly into the bowl's feed hole.

Because this 4160 carburetor has a single fuel inlet at the primary bowl only, an external fuel transfer tube connects the primary bowl to the secondary bowl. A rubber seal is used at each end of the transfer tube.

you wish to plumb the fuel line to the bowls, you can either select a fuel log with ends that thread into the adapter fittings or that have 7/8-20 ends that thread directly into the fuel bowls without the need for the original fittings.

A Holley banjo fuel inlet bolt requires an 11/16-inch wrench. The majority of 4150 and 4500 carb fuel bowl inlet fittings require a 1-inch wrench, but some carbs, such as the Ultra series, require a 3/4-inch wrench (same bowl thread size, but a slightly smaller and easier-to-access hex). The 1-inch hex fittings provide less-than-ideal access for a wrench because the hex is very close to the fuel bowl outer wall. The smaller hex allows much easier access.

This 4150 Ultra Double Pumper carb is equipped with an electric choke. Mechanical secondary operation is accomplished via a throttle shaft that connects one side to the other side. A lever connected to the secondary throttle shaft contacts and activates the secondary accelerator pump lever.

Vacuum-operated secondary control opens the secondary throttle plates based on engine requirements in terms of vacuum and load. The secondary throttle plates open only when the engine needs additional air and fuel, making a vacuum secondary design more easily adaptable to street applications.

The secondary vacuum assembly is connected to the secondary throttle shaft. As the engine demands, the diaphragm inside the assembly is drawn upward, overcoming the internal spring, and begins to pull the secondary throttle plates open.

As the engine demands additional air, the secondary throttle plates open. Even at wide-open throttle (WOT), the secondary plates only open to the full-open position if the engine is able to use this amount of air. This eliminates stumbling at WOT that might occur with mechanical secondary operation if the driver opens the secondaries before the engine requires the additional boost.

Mechanical versus Vacuum Secondary Operation

To avoid a stumble when the secondary plates open, the secondary plates must open at a slightly slower rate than the primary plates. This is true with either mechanical or vacuum operation.

When the primary throttle plates open to approximately 45 degrees, the secondary plates begin to open. The secondary opening rate must be quicker than the primary rate so that both reach WOT at the same time; it provides a smooth transition as engine speed and fuel demand increases. This transfer from primary to secondary throttle plate movement is called progressive movement. If the pedal is quickly and aggressively stepped on and the secondaries open before enough fuel is pulled through the venturis, the secondary fuel bowl's accelerator pump provides a quick shot of fuel to ease the transition, avoiding a stumble.

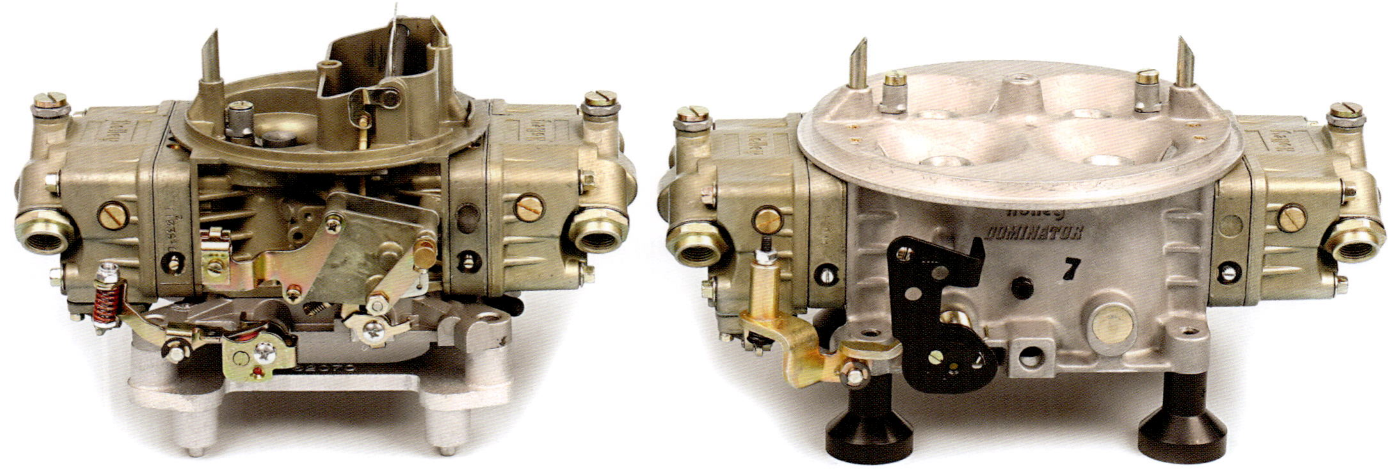

A 4150 with choke (left) and 4500 Dominator without choke provision (right).

Vacuum secondary operation, which is much more suitable to a street-driven application, controls the secondary plate opening via a diaphragm-equipped vacuum unit mounted on the passenger's side of the carburetor. The vacuum system opens the secondary throttle plates according to engine demand and is not dictated by the driver's right foot (that is, not mechanically). A spring inside the vacuum unit, on top of the diaphragm, keeps the diaphragm closed.

As the airflow through the primaries reaches a certain point and as engine speed and venturi vacuum increases, a vacuum signal from the primary venturi is routed to the vacuum unit, pulling the diaphragm open and overcoming the spring. When additional airflow begins to build in the secondary side, additional vacuum pull forces the diaphragm to open farther, opening the secondary plates farther.

Manual and Electric Choke

Choke operation is located on the passenger's side of the carburetor. If the carb is equipped with a manual choke, a control cable is required to allow manual choke

An electric choke is powered by a 12-volt signal. The voltage signal must be connected to an ignition-on circuit only. Do not connect the choke to a coil positive terminal.

plate position control from the driver's area. An electric choke automatically controls the position of the choke plate when the engine is started. The temperature-variable flat-wound spring inside the choke housing contracts and relaxes based on temperature, controlling the choke plate position. HP-series carbs (including the 4500 Dominator series) have no choke provision, and are designed for optimum performance. The elimination of a choke

horn and choke plate increases airflow to the venturis.

A choke system is needed for street operation for which the driver expects the engine to start quickly and attain an acceptable idle when the engine is cold. When the metal surfaces inside the carburetor are cold, fuel doesn't atomize readily. Also, the starting attempt occurs when there is little vacuum signal, which means that there is little vacuum draw to pull fuel out of the bowl. A choke plate positioned

The choke housing has a reference mark. By loosening the three screws that secure the black choke housing, the housing may be rotated to adjust the rate of choke operation. Rotating the housing clockwise richens it by keeping the choke plate closed, delaying its opening. Rotating the choke housing counterclockwise allows the choke to open sooner.

The choke housing includes a temperature-sensitive spring. The spring has an eyelet end that engages onto the internal choke lever tang.

above the primary venturis is closed to increase the vacuum signal, which helps to draw fuel out of the idle and main circuits. With less air and more fuel, a slightly rich mixture is created that helps the engine to start and run until the metal surfaces and the intake manifold reach an acceptable rise in temperature, at which time the choke plate is opened.

In warm ambient temperatures, a choke may not be needed at all. A few pumps of the throttle pedal causes the accelerator pump to send fuel into the venturis, allowing the engine to start, although the engine may idle/run a bit rough until the operating temperature rises to the point where fuel is atomized.

With a manual choke, the driver has immediate control over the position of the choke plate. With an electric choke system, specifically the integrated choke system on Holley carbs, the choke plate is held in the closed position when the heating element inside the choke housing is cold. The choke housing is connected to a 12-volt ignition circuit source.

A manual choke system is operated by the driver via a cable connection to the choke lever. In the full-open position (choke off), the lever is pushed forward.

This choke plate is closed (choke on) with the choke lever pulled rearward. A manual choke allows the driver to maintain full control of the choke plate position.

Main Jet Sizes

Jet Number (stamped)	Drill Bit Size (inch)	Jet Number (stamped)	Drill Bit Size (inch)	Jet Number (stamped)	Drill Bit Size (inch)
40	.040	61	.060	81	.093
41	.041	62	.061	82	.093
42	.042	63	.062	83	.094
43	.043	64	.064	84	.099
44	.044	65	.065	85	.100
45	.045	66	.066	86	.101
47	.047	67	.068	87	.103
48	.048	68	.069	88	.104
49	.048	69	.070	89	.104
50	.049	70	.073	90	.104
51	.050	71	.076	91	.105
52	.052	72	.079	92	.105
53	.052	73	.079	93	.105
54	.053	74	.081	94	.108
55	.054	75	.082	95	.118
56	.055	76	.084	96	.118
57	.056	77	.086	97	.125
58	.057	78	.089	98	.125
59	.058	79	.091	99	.125
60	.060	80	.093	100	.128

Obtain the 12-volt power from the ignition circuit only. Do not take 12-volt power from the coil. When the engine is started (aided by the closed choke plate), the electrical heating element begins to warm. As the temperature inside the heating element increases, the choke plate begins to open and attains full-open position after the heating element reaches its designated temperature. A vacuum tie-in between the choke base housing provides air to prevent the choke's heating element from overheating.

The electric choke is adjustable by loosening the three screws that secure the plastic choke housing to the choke base. The plastic choke-housing cap is rotated clockwise or counterclockwise to adjust choke operation (caps are marked "rich/lean" with arrows indicating which direction richens or leans).

4150 Vacuum and Mechanical Secondary Carbs

The 4150 has a dual-feed center-hung fuel bowl at both the primary and secondary sides. When equipped with vacuum secondary, no accelerator pump is located at the secondary side. A vacuum unit mounted to the passenger's side controls the secondary throttle shaft. The baseplate (also called the throttle body) has a large vacuum port for positive crankcase ventilation (PCV) operation and a smaller vacuum port for accessory use.

Many 4150s also have a second large vacuum port at the rear for power brake booster operation. On the passenger's side of the primary metering block is another small vacuum port for timed/spark advance. At idle, this port has little or no vacuum signal, which prevents the distributor from advancing prematurely. If your distributor has mechanical advance, this vacuum port must be plugged.

A curb-idle adjusting screw is located on the driver's side. It allows you to set engine idle speed by opening and closing the primary throttle plate to control the amount of air entering the primary venturis.

On the passenger's side, a fast-idle cam adjusting screw is also included on models with electric choke.

The 4150-series carbs have jet-equipped metering blocks at both the primary and secondary sides. They are available in a variety of configurations, with mechanical or vacuum secondary operation, manual or electric choke, or no choke provision at all.

The main body has a choke horn and choke plate to control airflow during engine warmup.

Under the choke plate are the two primary venturis and boosters, as well as an accelerator pump discharge nozzle, or squirter. On models equipped with vacuum secondaries, there is no secondary accelerator pump or discharge nozzle.

The primary and secondary sides each have four small holes at the top that serve as air bleeds. The two inboard air bleeds are for the main circuit and the two outboard holes are for the idle circuit.

The primary metering block has two jets to control normal fuel flow and a power valve that acts as an auxiliary fuel supply during acceleration. Idle mixture is adjusted via a mixture adjustment screw on each side of the metering block.

All 4150 carbs have dual-feed center-hung fuel bowls. The floats are secured with center-mounted hinges. Each float presses on a needle and seat assembly. The float and needle/seat work together to control the fuel level in each bowl. On the top of each bowl is a nut that is used to adjust float level.

An accelerator pump arm is attached to the baseplate on the bottom driver's side of the primary side. This arm presses against the accelerator pump's diaphragm. The pump is activated every time the throttle is pressed; it sends fuel to the discharge nozzle immediately. A 4150 equipped with vacuum secondary operation does not have an accelerator pump on the secondary side. In fact, secondary accelerator pumps are only found on carbs with mechanical secondary operation.

On a carb equipped with a vacuum-operated secondary, the vacuum unit handles control of opening and closing the secondary throttle plate. As the engine runs and venturi air velocity increases the vacuum signal in the venturi, the vacuum assembly automatically starts to open the secondary throttle plates when needed to supply the proper amount of fuel. To fine-tune secondary operation, you can change the vacuum assembly spring to a spring of a different rate.

A connecting link ties the rear throttle plates directly to the primary throttle lever. The 4150 Double Pumper with mechanical secondary operation allows you to manually control the opening and closing of the primary and secondary plates as you operate the throttle lever.

Secondary Metering Plate Jet Sizes

Stamped PN	Holley PN	Main Jet Orifice (inch)	Idle Orifice (inch)	Stamped PN	Holley PN	Main Jet Orifice (inch)	Idle Orifice (inch)
18	134-18	.064	.028	22	134-22	.076	.028
30	134-30	.064	.029	24	134-24	.079	.035
13	134-13	.064	.031	44	134-44	.081	.029
33	134-33	.064	.043	49	134-49	.081	.033
8	134-8	.067	.026	21	134-21	.081	.040
23	134-23	.067	.028	31	134-31	.081	.052
16	134-16	.067	.029	29	134-29	.081	.063
9	134-9	.067	.031	46	134-46	.082	.031
36	134-36	.067	.035	25	134-25	.086	.043
6	134-6	.070	.026	47	134-47	.089	.031
19	134-19	.070	.028	55	134-55	.089	.037
20	134-20	.070	.031	27	134-27	.089	.040
41	134-41	.070	.033	26	134-26	.089	.043
35	134-35	.071	.029	54	134-54	.093	.040
39	134-39	.073	.029	15	134-15	.094	.070
37	134-37	.073	.031	50	134-50	.096	.031
17	134-17	.073	.040	45	134-45	.096	.040
10	134-10	.076	.026	14	134-14	.098	.070
42	134-42	.133	.026				

CARBURETOR SELECTION AND SETUP TIPS

Selecting the right carburetor for a given engine and vehicle combination can seem confusing. The first order of business is to choose the appropriate airflow capacity, or CFM (cubic feet per minute), especially if your goal is to maximize horsepower. Remember the old adage, "bigger is not always better." If a carburetor is too small for a given application, it doesn't provide enough air and fuel for the engine to reach its potential. By the same token, if the carb size (CFM rating) is too great, the engine will likely bog, which also prevents maximum performance.

Next, consider the application, or how the vehicle will be operated. If the vehicle is intended strictly for the street, a carburetor equipped with vacuum secondary operation is almost always better than a double-pumper carb with mechanical secondaries. Although a mechanical-secondary carburetor can be tuned for street operation, it's simply easier to go with one that's equipped with a vacuum secondary system; the secondaries open only when the engine's load requires their operation.

You must decide whether you need a choke for your application. If the vehicle is intended for the street, even if driven only occasionally, you need a choke. Restricting airflow to the primary side of the carb aids in engine startup, especially in cooler temperatures. If no choke is present, you need to adjust the throttle during startup, and then allow the engine time to warm before venturing onto public roads.

For a vehicle that is used only on the track (drag strip, road course, oval track, off-road competition, etc.), easy startup when the engine is cold isn't that critical, and you usually have adequate time for the engine to warm to operating temperature before making a run on the strip or before the first lap on a course. For a competition vehicle, eliminating a choke mechanism reduces weight, simplifies the makeup of the carb,

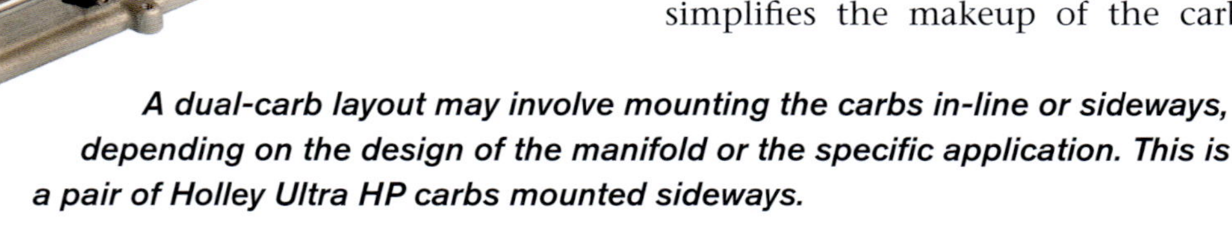

A dual-carb layout may involve mounting the carbs in-line or sideways, depending on the design of the manifold or the specific application. This is a pair of Holley Ultra HP carbs mounted sideways.

and eliminates any potential for primary airflow restriction if the choke plate sticks or binds.

If you like the idea of having a choke, consider the type of choke you prefer. A mechanical choke allows the driver to control the position of the choke plate via a manually operated cable. It provides total control; but the driver must be alert enough to open the choke when the engine no longer requires primary-side air restriction. Properly adjusted, an electric choke automatically restricts airflow and produces a richer fuel condition. After the ignition is energized, the choke plate automatically begins to open as the engine temperature rises. The choke plate moves to the full-open position as the engine warms to operating temperature and no longer requires this temporary rich condition.

Different carburetor models have varying numbers and locations of vacuum ports. If you're running power brakes, you need a full-manifold vacuum port to provide vacuum assist to the power brake booster. This may involve a port at the rear base of the carburetor to provide full manifold vacuum; the intake manifold itself may have a threaded port at the rear wall of the plenum, just below the carburetor-mounting flange.

Selecting Carburetor Size

Carburetor "size" (or CFM capability) is selected with a host of variables in mind, not just for the engine but also for the vehicle, drivetrain, and vehicle usage. If the vehicle is equipped with an automatic transmission, for example, consider the torque converter stall speed, which represents the lowest engine RPM at which WOT is applied. If the vehicle

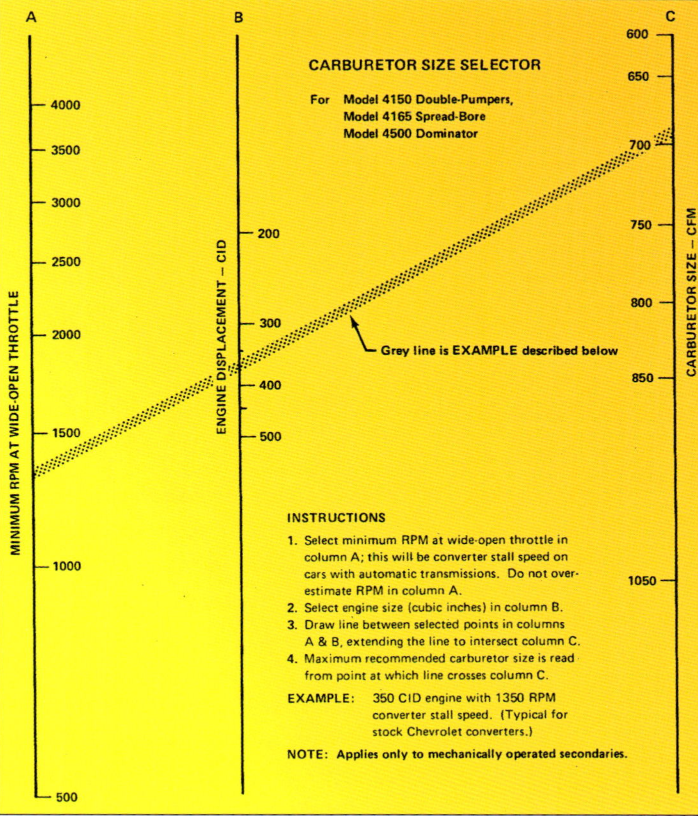

This carburetor selector chart aids in choosing carb size. Draw a line from the minimum engine RPM at WOT (the value range at the far left) through the value line for engine displacement. Continue this line to the carburetor size range at the far right. In this example, if the minimum RPM at which you expect to use WOT is about 1,350, and the engine size is 350 ci, the maximum suggested carburetor size is just a tick below 700 cfm. In this case, a 650- to 700-cfm carb should be adequate. Use this as a baseline; you still need to consider vehicle weight, axle ratio, and vehicle use. (Photo Courtesy Holley)

is equipped with a manual transmission, consider the lowest engine RPM that you expect to use at WOT.

The relationship between the lowest engine RPM that you expect to use at WOT and engine displacement roughly determines the most appropriate carburetor size.

Vehicle weight and axle ratio need to be factored into the selection process as well. A heavier vehicle and/or a numerically lower axle ratio typically prefer a smaller carburetor; a lighter vehicle and/or a higher axle ratio may prefer a larger carburetor. A lighter car, such as a Mustang or Camaro that is equipped with a numerically higher axle ratio, may require a slightly larger carb, on the order of 700 to 750 cfm.

Bear in mind that this applies when selecting a carburetor with mechanical secondary operation. If you plan to use a vacuum-secondary carb, you can always increase CFM; the secondary operation is enabled according to engine demand.

The carburetor must be matched to the application. Although it may be tempting to purchase a larger carburetor than the engine requires, you need to avoid it because it won't yield more horsepower. Most want maximum horsepower with fuel efficiency, and if your engine receives more fuel than it can effectively burn, the unburned fuel is pumped out the exhaust port and makes the spark plugs wet. Therefore, gas mileage suffers.

With a carb size that is too large, absolute top-end power might improve, but off-the-line and cruising mid-range acceleration can suffer (bogging). If you err on the side of caution by choosing a slightly smaller carb, you may sacrifice a bit of top-end power but your low-end and mid-range acceleration improves.

A formula to use as a base reference follows. Bear in mind that this formula assumes that the engine operates at 100-percent efficiency, which doesn't happen in the real world.

$$\text{Carburetor CFM} = \text{engine ci} \div 2 \times \text{maximum RPM} \div 1{,}728$$

Where:
1,728 = factor to convert cubic inches (ci) to cubic feet (cf)

This formula indicates half of the specified engine displacement because a four-cycle engine intakes once with each crank revolution. If the engine size is 455 ci and the maximum expected engine speed is 6,000 rpm, CFM is 789.93 (455 ÷ 2 x 6,000 ÷ 1,728).

In this example, the engine prefers the closest-available carburetor size of 800 cfm. Because most engines (assuming that it is properly built and is operating at its best) have a volumetric efficiency in the 80- to 95-percent range, a slightly smaller carb (in the range of 650 to 750 cfm) is a more realistic choice for this engine.

Before you try to determine the ideal carburetor size for a given application, you need to have an idea of the engine's volumetric efficiency (VE), which is an indication of the engine's ability to breathe. The better

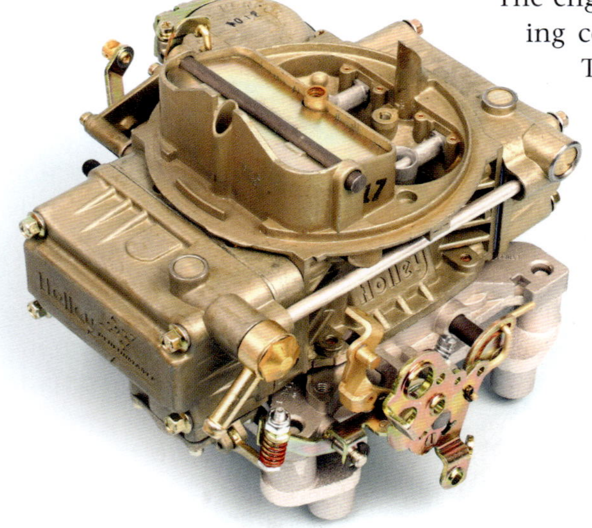

Although it may be tempting to choose a Holley Dominator by its reputation, super-cool looks, and hefty physical size, remember that it's only available in certain sizes (many of which may be too large for a street application). Also, there's no choke provision, so it's not going to be cold weather/street friendly. A Dominator has a unique mounting base, so a Dominator-specific intake manifold is needed (or an adapter that allows mounting a Dominator to a standard intake manifold). Dominators are designed for competition use.

Holley's Avenger series of carburetors are available in CFM sizes ranging from 470 to 870. These are ideal street carbs, featuring vacuum secondary operation and electric chokes.

the engine breathes, the higher the VE. This is expressed as a ratio of the air mass (weight of the air) that the engine intakes compared to the air mass that the engine's displacement theoretically ingests if there were no losses. VE is fairly low at idle and varies according to engine speed. VE should be considered according to the engine RPM at which the engine realistically operates.

An engine cannot operate at 100-percent VE in the real world; it simply isn't possible. To maximize VE, you need to verify some things. The engine must be in good operating condition and not worn out. This means that the compression ratio must be close to new, oil pressure is within acceptable ranges, all bearings operate well, and no other significant problems are evident.

The ever-popular 600-cfm 4160 (PN 1850) carbs have vacuum secondary operation and single fuel-inlet design, ideal for either single- or dual-carb installations.

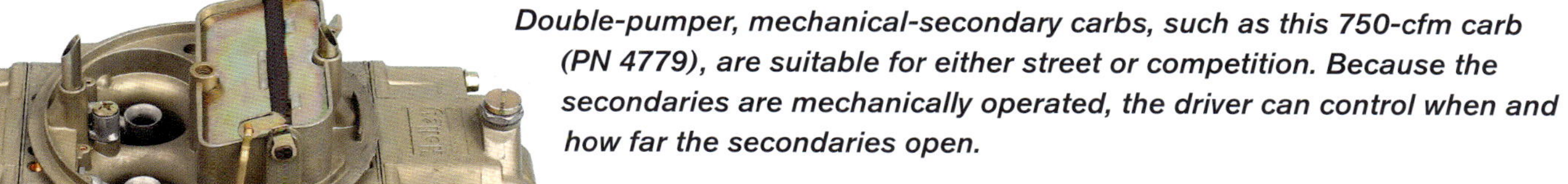

Double-pumper, mechanical-secondary carbs, such as this 750-cfm carb (PN 4779), are suitable for either street or competition. Because the secondaries are mechanically operated, the driver can control when and how far the secondaries open.

A typical low-performance engine may have a compression ratio in the range of 70- to 80-percent VE at the engine's maximum torque level. A typical high-performance engine that offers increased breathing may have a VE in the range of 80 to 85 percent at maximum torque. A top-level racing engine may have a VE in the range of 90 to 95 percent at maximum torque RPM.

The variables of engine parameters, including tuning the intake and exhaust system, cylinder head porting, camshaft profile, etc. all work in combination to increase VE. You can consult one of Holley's carb selection charts. Based on 100-percent VE, if the chart indicates that the best size is, for example, 700 cfm, if the engine has around 85-percent VE. So, in theory, a better choice for that engine might be a carb size of 600 cfm.

Forced induction (such as supercharging) usually requires about 40- to 50-percent more carburetor capacity compared to a naturally aspirated engine. (Holley now offers a line of carburetors designed specifically for supercharged applications.)

Charts and formulas aside, consider moving to a bigger carb when displacement increases, when higher RPMs are to be used, when engine compression is higher, when mechanical ignition advance increases, when higher-ratio drive gears are used, and as vehicle weight is decreased. A smaller carb suits conditions when engine compression is low, the vehicle is heavier, when torque converter stall speed (on an automatic transmission) is very low, and when the drive gear ratio is numerically lower.

Also, keep in mind that a given carb (or carb tuning) used during an engine dyno session doesn't necessarily perform the same when the engine is in the vehicle. The engine dyno run is merely a starting point; additional tuning (and possibly a move to a different carburetor) is potentially necessary once the vehicle is on the track (or road).

Carburetor Linkage

The throttle may be actuated via the accelerator pedal by either a rod linkage system or a flexible throttle cable, depending on the specific vehicle or the direction chosen by a custom vehicle builder. If a throttle cable is employed, it must be routed as smoothly as possible, with no sharp bends or kinks. A kinked cable or extreme bends can easily result in throttle binding, and this can lead to a sticking throttle at high engine speeds, which is an obvious safety hazard, to say the least. Also, make sure that a throttle cable is kept far enough away from any extreme heat source, such as the exhaust manifold or headers.

Carefully examine any throttle cable before installation to verify that it moves easily and freely within its casing. Whenever a throttle cable's condition is questionable, it's best to just replace it with a new cable. A throttle cable's casing must be anchored at a convenient point to prevent the entire cable assembly from moving during acceleration and deceleration. Only the internal cable must be able to move, not the exterior sheathing.

A cable retention bracket should be mounted either on the driver-side rear carburetor mounting bolt at the intake manifold or on the

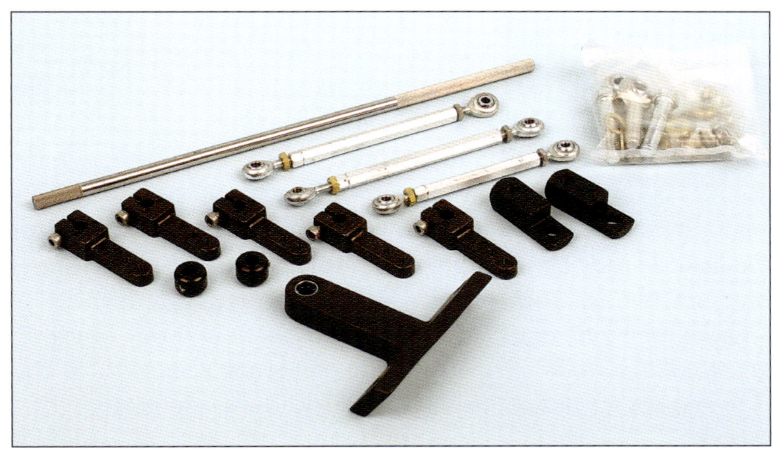

Multiple-carb throttle linkage systems are available to allow control of a dual-carb setup.

Adjustable-length rods equipped with radiused rod ends provide throttle adjustment and reduced friction. Due to multiple pivot points, rods must be adjusted carefully to allow sufficient play to avoid binding and to provide throttle activation at WOT.

A bellcrank is the pivot base where the throttle cable (or rod) connects and transfers movement to the carburetor linkage. This must be mounted rigidly, as shown on this custom build; the base is secured to the intake manifold.

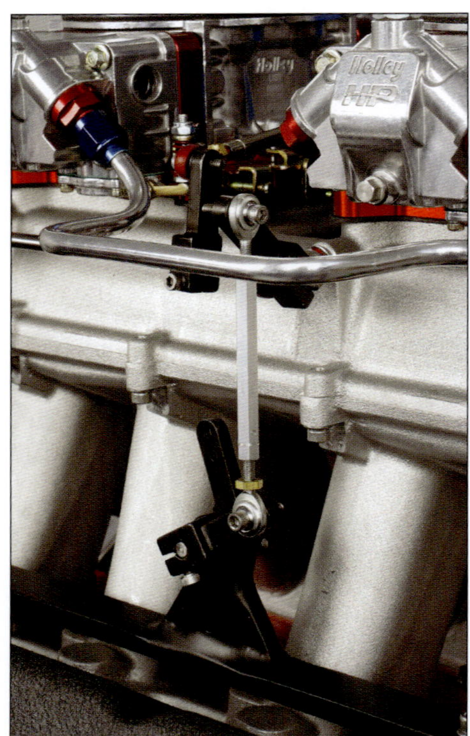

Make sure that the linkage rods do not interfere with adjacent areas such as fuel lines, the intake manifold, etc. Manually operate the linkage from idle through WOT to inspect for possible interference areas.

Here, the throttle linkage is being adjusted on a 1968 Mopar, using the original throttle cable and transmission kickdown linkage rod system. Note the return spring that pulls the throttle closed. Certain automatic transmissions, such as the Chrysler TorqueFlite, require careful rod adjustment at the carburetor throttle lever.

intake manifold itself. A variety of cable mounting brackets is readily available to suit any carb/manifold application.

If the throttle actuation is handled by a linkage rod system (1960s and 1970s Mopars are a good example), the linkage must be assembled correctly and all pivot points must be able to move freely without binding. You must pay strict attention to the assembly procedure, especially when dealing with a linkage rod assembly (rather than a throttle cable); make sure that the rods do not interfere with any nearby surfaces.

Carefully monitor the linkage system, from closed throttle to WOT. Make sure that the linkage does not touch any other surfaces that will cause the linkage to bind or rub. This includes the cylinder head, valve-cover, rear of the block, any nearby screws, bolt heads, hoses, wiring, etc.

Pay attention to linkage rod geometry. Linkage rods (or cable connections) should be set up to push or pull the throttle lever in a straight line. When viewed from the side, they should be parallel with the base of the carburetor. When viewed from above, the linkage rods or cable should be straight and parallel with the side of the carburetor. Misalignment (when the rod or cable is at an offset angle) can result in excessive side loading and can require more force to operate the throttle. This can lead to premature wear of the carburetor's throttle shaft.

Adjust the linkage to prevent "over-pull" or "over-push" at the throttle lever. At WOT, the linkage should be adjusted so that the force applied does not try to push the throttle lever beyond its WOT point. If this limit-stop is not adjusted properly, a heavy foot applied to the throttle pedal exerts excess force at the throttle lever, potentially causing premature throttle shaft wear and possible distortion of the throttle lever.

Do not rely solely on the carburetor's built-in throttle-lever return spring for throttle closing. An additional return spring must be used to ensure that the throttle closes in real time relative to the operation of the throttle pedal. Dual springs with an inner and outer spring are popular choices. The lighter inner spring applies a light tension that assists throttle closing; the heavier outer spring provides added muscle to pull the throttle lever closed.

Installation of the spring is similar to the linkage installation. It should provide a straight pull, with the spring installed as close to parallel to the carb baseplate as is possible. With the engine off, test the throttle while monitoring the spring to make sure that the spring does not rub,

interfere, or hang up at any point from closed to WOT.

Carefully adjust and test all throttle lever connections, including cable and/or linkage, and the return spring(s) before firing the engine. Never start the engine until you are absolutely certain that you have no binding issues, and that the throttle closes easily. Performing these initial inspections with the engine running poses the very real risk of the engine running wild at WOT, with the potential for a disastrous engine failure.

Vacuum Fittings

Carburetors that are equipped with a vacuum port to accommodate a vacuum-advance distributor have a small-diameter vacuum fitting (requiring a 1/8-inch hose) located on the passenger's side of the primary metering block, just above the mixture screw. If you have a distributor with vacuum advance, connect the distributor vacuum hose to this timed port on the carburetor to eliminate spark advance at closed

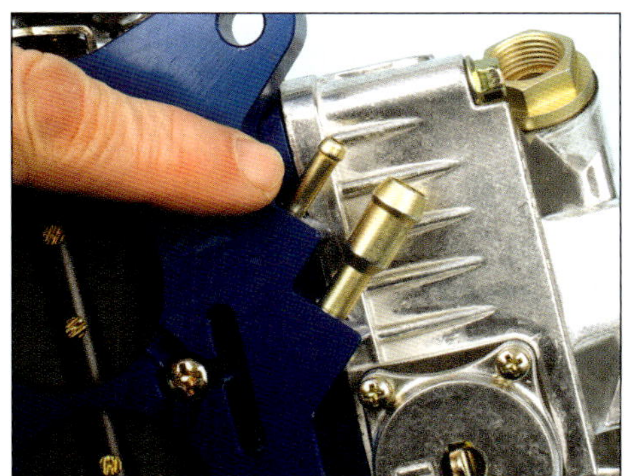

This view from the underside of the baseplate shows the front-mounted vacuum fittings. The larger 3/8-inch fitting is intended for PCV hose connection. The smaller 1/8-inch fitting is for various accessory components' vacuum. Both provide full manifold vacuum.

The 1/8-inch timed vacuum fitting for connection to a vacuum advance distributor is located on the passenger's side of the primary metering block. If you don't need vacuum advance, plug this fitting with a rubber cap.

throttle. A port in the throttle bore is exposed to vacuum as the throttle plate moves past the port, usually at slightly off-idle, providing the vacuum signal to the distributor.

A large-diameter full-manifold vacuum port that requires a 3/8-inch vacuum hose (to provide a vacuum signal to a power brake booster) is located at the rear of the carburetor baseplate (depending on the carburetor model). If this is not included on

This view from the underside of the baseplate shows the rear 3/8-inch vacuum fitting for power brake booster vacuum.

This overhead view reveals the two full-time manifold vacuum fittings located on the front of the baseplate.

This view from the rear shows the 3/8-inch-diameter vacuum fitting for power brake booster connection.

your carb, full-manifold vacuum can be accessed at the rear of the intake manifold plenum.

If neither the carb nor the intake manifold has a full-vacuum port at the rear, you can drill and tap the intake manifold plenum to accept a 3/8-inch NPT fitting to provide a full-vacuum signal to the power brake booster. Of course, this should be performed with the intake manifold removed from the engine. All current Holley carburetors, with the exception of the 4500 Dominator series, already have this large-diameter full-manifold vacuum fitting at the rear of the baseplate.

Some carb models also have two additional full-time vacuum port fittings at the front of the baseplate. They also include a 3/8-inch fitting for PCV hose connection and a 1/8-inch fitting. These provide full vacuum for operation of the air cleaner, air pump diverter valve, A/C, cruise control, and, if the vehicle is so equipped, temperature-sensing valve.

Any vacuum port fitting on the carburetor that is not needed must be closed off with a rubber vacuum cap. Carburetors such as the 4150 (including Avenger and Ultra) have four vacuum fittings. These are

for the full-time rear power brake booster, the full-time front PCV, the full-time front accessories, and the timed vacuum on the passenger's side of the primary metering block for the vacuum distributor advance. The PCV vacuum fitting is located on the front (primary side) of the baseplate because the primary usually flows more air, and also so that the crankcase vapors pulled by the PCV are shared evenly between the two primary throttle bores.

Chokes

If your carb is equipped with an electric choke system, adjustments are available that allow you to set the opening and closing time/duration of the choke plate. Initially, the choke should be open enough to provide throttle response without engine stalling. If the choke comes off too early (doesn't stay on long enough), rotate the choke's thermostatic choke cap (the black housing) counterclockwise. If the choke comes off too late (stays on too long), rotate the choke cap clockwise. Always rotate the cap one notch at a time and test the operation until it is satisfactory. Be sure to re-tighten the three choke cap hold-down screws.

A choke that comes off too early can result in engine stalling, surging, stumbling, or backfiring when the engine is cold. A choke that comes off too late causes an over-rich condition, resulting in rough idle, poor driveability, lousy fuel mileage, and black smoke emitted from the exhaust.

The choke's thermostatic spring cap is marked on the lower face to indicate which direction to rotate the cap for rich or lean adjustment.

As a reference point, note the original location before loosening and adjusting the choke cap. Then, note the small groove in the black cap; the adjustment mark on the choke cast housing that aligns to this groove. When turning the cap for choke adjustment, move the cap one mark at a time and monitor choke operation. You may want to color the mark on the casting that aligns with the cap's groove with a Sharpie, to provide an easy-to-see reference.

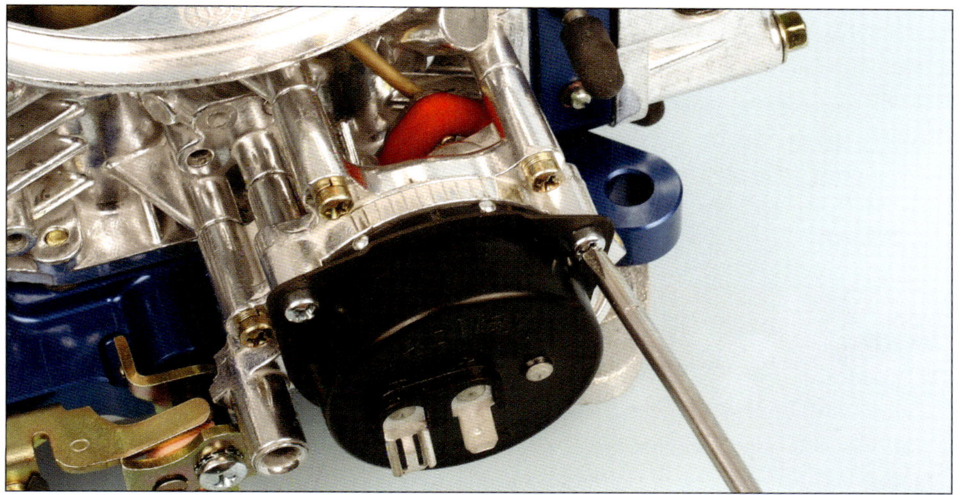

To adjust choke opening and closing points, loosen the three screws that secure the choke thermostatic housing cap with a Phillips screwdriver. When adjusting the cap position, perform adjustments one reference mark at a time.

The fast-idle cam is found on the rear of the choke housing. The plastic cam includes a series of steps that contact the fast-idle cam adjusting screw pad. On initial cold start, the cam's highest step contacts the adjuster screw arm's pad. As the engine warms, the cam progressively steps down until the choke plate is fully open.

The choke mechanism has a fast-idle cam that is connected to the choke thermostat. The cam has a series of steps that allow idle to step down during warm-up until the choke plate is fully open; it contacts an adjustable screw on a lever that connects to the primary throttle plates. The throttle stops against this plastic cam to raise engine speed during choke operation; holding the choke at a greater throttle angle produces a higher idle speed. The higher idle speed aids in atomizing the fuel for cold-engine operation. Before starting a cold

Here's a close-up view of the fast-idle cam adjuster screw lever. Note the nylon pad at the tip of the arm. This pad contacts the fast-idle cam steps.

To adjust the fast idle, the engine must be off and cooled; open the throttle to WOT and hold it open. This moves the fast-idle lever down and allows easy access to the screw. The screw can be adjusted using an open-end 1/4-inch wrench.

engine, hold the throttle pedal to the floor; the fast-idle cam holds the primary throttle plates open more than usual, to raise idle RPM during initial start and warm-up. The fast-idle speed is set at the factory, and should provide 1,500 to 1,600 rpm. If idle speed requires adjustment, simply adjust the fast-idle screw.

With the engine off, allow the engine to completely cool to ambient temperature, making sure that the choke plate has fully closed. With the engine off, manually open the carb to WOT and hold it there. This exposes the fast-idle screw (located behind and to the lower rear of the choke housing). Using a 1/4-inch open-end wrench, turn the screw clockwise to increase RPM or counterclockwise to decrease RPM.

Close the throttle, start the engine, and check engine speed. Re-adjust until you achieve approximately 1,500 to 1,600 rpm at cold fast idle. As the engine warms, the cam drops onto its second step as the vacuum choke pull-off opens the choke slightly. The cam eventually drops to its last step when the choke plate opens fully.

If the carburetor is equipped with an electric choke, connect the choke cap's positive terminal directly to a steady 12-volt power source that is activated by the ignition switch. Do not connect it to the ignition coil. A suitable 12-volt connection includes leads or wiring for vehicle accessories, such as the windshield wiper motor. However, do not connect the choke's power lead to the original equipment electric choke source, because it might not be a 12-volt source.

Fast-Idle Screw Clearance

Depending on the intake manifold, pay attention to the clearance between the fast-idle screw and the manifold when the screw's adjusting arm is held down under WOT. If the screw contacts the manifold surface, you may need to do one of two things for clearance: install a carburetor spacer or grind a slight relief in the intake manifold. ■

The choke cap has two terminals. Each is marked on the cap for negative and positive (indicated) connections. The positive power terminal must be connected to a 12-volt positive power source that activates with the ignition in the ON position. Do not connect the choke's positive to the ignition coil's positive post.

TOOLS AND EQUIPMENT

Carburetor service requires a selection of conventional hand tools as well as a select number of specialty tools. Many specialty tools are available from Holley as well as popular aftermarket suppliers, such as Summit Racing and Jegs. Having all of the tools listed here prior to beginning service makes your work more efficient and less time-consuming. When a specialty tool is required to perform a procedure, don't substitute a general tool because you may damage the components and parts. Before you begin, clean your workbench surface and make enough room to spread out all parts in an organized manner.

Basic Hand Tools

To completely disassemble and reassemble your Holley carb, you need a variety of hand tools. Carburetors are sensitive components and made of aluminum so you need to use the right tool for the job. In most cases, jets, screws, and other fasteners are installed with a light amount of torque, so be cautious and patient. Don't force or overtorque any fasteners on the carburetor because you can easily damage vital components.

The benefits of using high-quality hand tools should be obvious. Cheap, bargain-basement tools may not be sized properly, and cheap chrome plating can flake off and contaminate your work. Although it's certainly not necessary to buy the most expensive tools available, try to avoid extremely cheap, poor-quality offshore tools that may not have the required manufacturing tolerances and/or may not be hardened properly. Better-quality tools last longer, fit better, and reduce the potential for avoidable frustration.

Before disassembling a carburetor, take the time to clean your tools. Granted, you need to clean the parts anyway, but it just doesn't make sense to add dirt, grit, and grease when you can avoid it. And certainly, before reassembly, it is absolutely

Resting a carburetor directly onto a workbench without raised fixtures can possibly result in bending linkages if excess downward pressure is applied during disassembly or assembly. Any carburetor should be properly secured and raised off of the work surface during any off-vehicle service.

Basic Hand Tools

Tool	Function
1/2-inch socket wrench, extension and ratchet	Carburetor removal from manifold (preferably 1/4-inch-drive to maximize access)
1/2-inch combination wrench	Carburetor removal (depending on application; 5/16-inch bolts or nuts are most common for carburetor-to-manifold mounting, which requires a 1/2-inch wrench, but mounting applications may vary)
Nut drivers	
3/8 inch	Fuel-bowl screws
Small sharp chisel, metal or plastic scrapers	Gasket removal
Razorblades	Gasket removal
Picks	O-ring removal
Phillips screwdrivers	
No. 1 tip	Accelerator pump cover screws
No. 2 tip	Throttle-plate screws
Flat-blade screwdrivers	
1/8-inch tip	Idle adjustment
3/16-inch tip	Pump-cam screw
1/4-inch tip	Fuel-bowl screws
Torx screwdriver	
T20	Secondary jet plate/choke covers (late models)
Torque wrench	
1/4-inch drive (in-lbs)	Torquing fasteners to specification (40 to 200 in-lbs)
Plastic/rubber mallet	Loosening bowls, etc.
Needle-nose pliers	Handling small parts
Combination wrenches	
1/4- to 1-inch	Disassembly and assembly
Fuel-line wrenches	
1/2-, 5/8-, and 9/16-inch	Fuel line fittings
Feeler gauge set	Checking clearances
Machinist's straightedge	Checking flatness
Fine flat file	Deburring edges
Notepad or laptop	Recording service notes (jet sizes, condition, disassembly order, etc.)
Lint-free paper towels	General cleaning and organizing parts on a clean surface
Safety gloves and goggles	Protection when using cleaning solvents
Compressed air can	Blowing small passages clean

Main jets have a slot that can be accommodated with a screwdriver, but a dedicated jet driver is preferred. Pictured here is Holley's PN 26-68. The jet service tool's handle is designed for a no-slip grip.

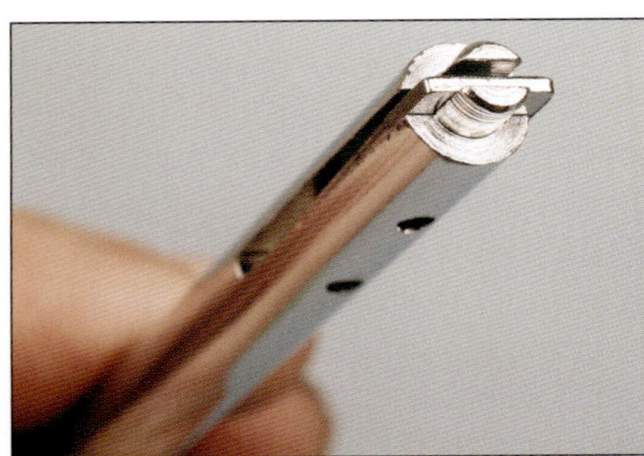

The driver tip includes a flat-blade that engages the jet, while a round pilot engages into the jet's recess for a secure fit, eliminating accidental slipping and potential burring of the jet.

A 5/16-inch nut driver tool makes fuel-bowl screw service quick. During initial tightening, the small grip area of the tool helps to avoid overtightening.

A machinist's precision straightedge can be used, along with feeler gauges, to measure carburetor baseplates for warpage. Don't rely on just any ruler. A machinist's straightedge is precision ground to provide a perfectly flat and straight edge.

If you plan to perform precision carburetor service, it's best to avoid guessing at fastener torque values and clamping loads. If you have the time, take advantage of a precision torque wrench to avoid leaks and mating surface distortion. A 1/4-inch drive torque wrench calibrated for 40 to 200 in-lbs is adequate. Do not use a torque wrench designed for foot-pound values!

Commonly, a 5/8-inch wrench and a flat-blade screwdriver are used to adjust floats. These are the tools that everyone has been using for decades, and there's nothing wrong with this approach. A dedicated adjuster tool is simply more efficient, especially for racers who may need to perform service more often or in a hurry.

limited to a quick exterior wipe-down. Properly clean all tool surfaces, including the inside of any socket, wrenches, ratchet or driver handles, screwdriver handles, shanks, and tips, etc. Also, take the time to inspect your tools for burrs or flaking plating.

Specialty Tools

Instead of using a flat-blade screwdriver for jet removal and installation, a dedicated jet removal tool is not only available but is highly recommended. An example is Holley's jet tool (PN 26-68). It is specifically designed for jet service, offering a no-slip secure connection to a jet. Plus, it's only used for jet servicing; it won't become burred or otherwise damaged as often happens to a conventional screwdriver. It will always be ready for proper engagement to a jet when necessary.

Power-valve removal/installation tools are available designed specifically for power-valve engagement. One example is from Willy's Car-

buretors (PN 3104). Because power valves have an internal diaphragm, diagnosing problems is easy with a simple power-valve tester tool. This tool has a removable head. Insert the power valve into the tool, install the head, and connect the head's vacuum port to a hand-held vacuum tester. Power-valve testers are available from a variety of manufacturers.

In addition, dedicated float/ needle and seat adjustment tools are available that incorporate a nut driver and screwdriver in one tool. A 5/8-inch open-end or box wrench and a flat-blade screwdriver are commonly used, but this type of specialty tool is handy and designed specifically for this task.

For enthusiasts and racers who routinely experiment with different jet sizes, a dedicated jet card is a necessity. Aluminum cards that have threaded holes to store an array of jets are extremely handy and eliminate sorting through a pile of jets stored in a can. These jet cards are available from a variety of sources, including AED (PN 6020).

When bench servicing a carburetor, using a stable fixture raises the carb above the workbench surface, which provides better access to various areas of the carb, and allows movement of the linkage. Carburetor work fixtures are available in three basic designs. The simplest (and least expensive) is a set of four plastic "legs," each of which snap into the carburetor baseplate's four mounting bolt holes. Another design is a one-piece plastic, aluminum, or steel platform with four raised pegs that engage

mandatory to make sure that all tools are *clean* to avoid introducing contaminants to a clean carburetor. Cleaning your tools shouldn't be

Here's a Holley Avenger carb mounted atop a set of plastic legs. This provides enough clearance to allow working the linkage without contacting the workbench surface.

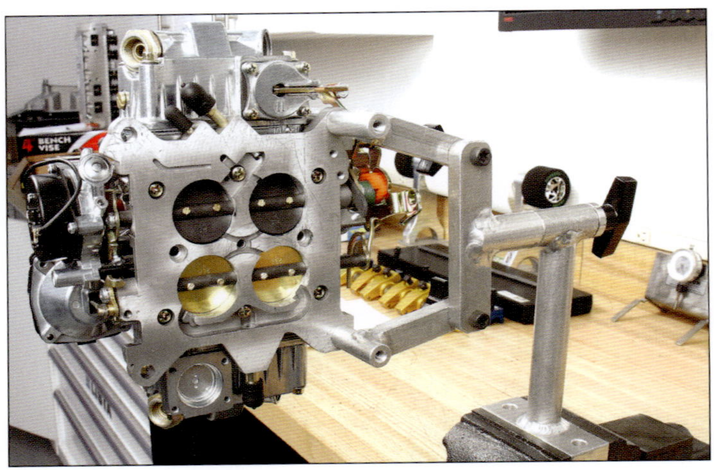

The AED carburetor work fixture (PN 7920) can be bench-mounted or secured to a bench vise. Two long T-handles secure the carb base to the fixture. The carb is positioned at a convenient working height. The AED carb fixture can be rotated 360 degrees for easy access to any service area. Here the carb is positioned 90 degrees.

This AED fixture, with the carb positioned upside-down, is solid and sturdy. The mounting arms pivot to align with one side of any carb's baseplate.

Moroso's cast-aluminum carb stand (PN 62070) has peg supports on each side, to accommodate either square-bore or spread-bore carburetors.

This is a very handy tool, especially for carburetor work. This example from Lisle is designed to pick up and hold small linkage rod clips.

When the thumb-activated spring-loaded head is pressed, a small angled probe protrudes from the tip, to engage a small clip. Once the clip has been grabbed, releasing the thumb button secures the clip, preventing it from flying out into no-man's land.

Although a 5/8-inch box wrench and a flat-blade screwdriver can certainly be used to adjust the float on a Holley carburetor, a handy option is to use a dedicated float adjuster tool, such as this one from Willy's Carburetors.

The head is installed, capturing the power valve. During testing, an open passage allows you to observe valve movement as vacuum is applied to the tool.

Specialty Tools

Tool	Function
Hemostat and tweezers	Handling small parts
Small pivoting mirror	Inspection
Small flashlight	Inspection
Magnifying glass	Inspection of small parts
Magnetic pick-up tools	Holding/retrieving small screws, clips
Jet driver	Jet removal and installation
Jet storage card	Organize spare jets
Carb workbench stand	Sturdy mount for workbench service
Dye penetrant	Crack checking
Power-valve tester	Check power valves (requires use of hand-held vacuum pump)
Carb pin tool	Service small clips
Power-valve removal tool	Removal and installation of power valve
Carb booster installation tool	Swedge installation of new booster
Jet cleaning kit	Small-diameter rod kit for cleaning jets
Float level gauge	Checking float height position
Float adjustment/needle and seat tool	Combined nut holder and lock-bolt tool (eliminates need for wrench and screwdriver)
Parts basket or pan	Safely holds carburetor parts during cleaning
Drill bits	Check or enlarge main jets, air bleeds, etc. (.040 to .156 inch)

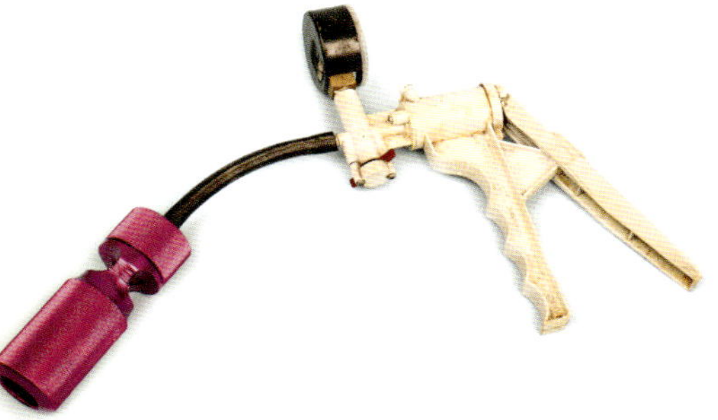

With the power valve installed in the tool, connect a hand-held vacuum pump and apply vacuum to test the power valve.

A jet card, made of aluminum and threaded to accept primary and secondary jets, is a must-have for anyone who experiments with jet sizes for performance tuning. This card, from AED, neatly stores and organizes as many as 44 jets. Dumping spare jets into an old coffee can simply isn't a good idea.

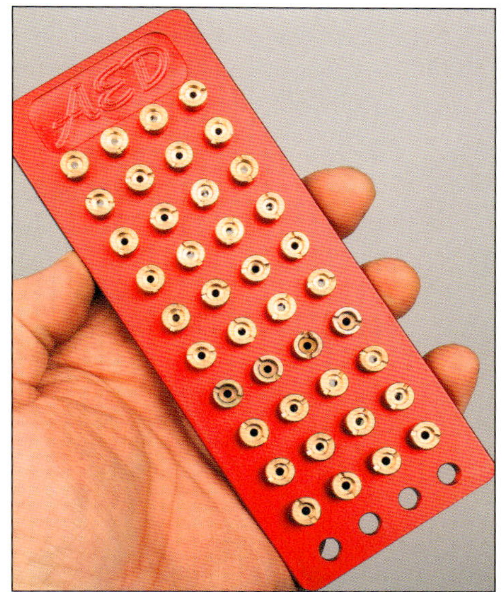

AED's booster installer (PN 6070) allows replacement boosters to be installed and swedged into place in the main body.

If annular boosters are to be installed, AED offers an annular booster adapter (PN 6-74) that mounts to their installer tool, to accept an annular-style booster.

into the baseplate bolt holes. A third design is more complex and can be secured in a vise or bolted to a work-bench. It allows the carb to be moved away from the workbench and allows you to pivot the carb during service.

Lubricant Compatibility

Some lubricants are not compatible with O-rings or other rubber materials. Avoid using silicone spray or some thin lubricants such as WD-40, which may not be compatible with rubber parts. ■

Drill Bits for Main Jet Sizes

Following is a listing of Holley main jet numbers and their specific drill bit sizes. Remember that the number that is stamped onto a jet's outer body does not necessarily directly indicate its orifice/drill bit diameter.

Jet Number	Drill Bit Size (inch)	Jet Number	Drill Bit Size (inch)
40	.040	71	.076
41	.041	72	.079
42	.042	73	.079
43	.043	74	.081
44	.044	75	.082
45	.045	76	.084
47	.047	77	.086
48	.048	78	.089
49	.048	79	.091
50	.049	80	.093
51	.050	81	.093
52	.052	82	.093
53	.052	83	.094
54	.053	84	.099
55	.054	85	.100
56	.055	86	.101
57	.056	87	.103
58	.057	88	.104
59	.058	89	.104
60	.060	90	.104
61	.060	91	.105
62	.061	92	.105
63	.062	93	.105
64	.064	94	.108
65	.065	95	.118
66	.066	96	.118
67	.068	97	.125
68	.069	98	.125
69	.070	99	.125
70	.073	100	.128

Chemicals

Several chemicals and lubes are useful during the assembly of a carburetor and its components. Have these lubes on hand so the components are assembled accurately and with minimal stress. Carburetor cleaning solvent (such as CRC TYME-1) is required, and a 1- to 5-gallon bucket for soaking parts is useful. A can of carb cleaner from Gunk or another suitable brand is also needed. Vase-line is used for lubricating O-rings. Lithium grease is suitable for throt-tle linkages and high-tack adhesive is good for securing gaskets in place.

Drill Bits

A selection of small-diameter drill bits accommodates modifica-tions (if you choose to modify and/or experiment). Examples include enlarging main jet sizes during test-ing to determine what size jets to purchase, slightly enlarging idle air bleeds to lean the idle mixture, etc. Bear in mind that before using any drill bit to modify components, you must carefully measure the shank's diameter with a micrometer to ver-ify the diameter of the bit. Drill bits can be unmarked, misplaced in their cases, etc. Never use a bit unless you first measure its diameter.

Torque Wrench

Carburetors are precision assem-blies. Undertightening or overtight-ening of various threaded fasteners can easily lead to leaks, mating sur-face warpage, and even cracking (see Chapter 7 for various torque values). "Guesstimate" tightening simply isn't a reliable method, so you need a quality calibrated torque wrench

Solvent Safety

Carburetor cleaning solvent is strong stuff, and you need to avoid skin contact. Always wear protective rubber gloves, a long-sleeve shirt, and safety glasses or goggles. In addition, a shop apron can help to protect your clothing. ■

designed for inch-pound values. Because the torque values involved in carburetor work are on the "light" side, a 1/4-inch-drive inch-pound torque wrench is the best choice. The most accurate styles include micrometer, click-type ratcheting, or dial torque wrenches. A torque wrench with a range of 40 to 200 in-lbs is adequate for carburetor work.

Torque wrenches must be handled and stored with care. They should be recalibrated about once each year, with service intervals based on wrench use. Torque wrenches that are used on a regular basis require recalibration more often. Recalibration services are relatively inexpensive and are available from specialty service shops and through the tool's manufacturer.

Cleaning Chemicals

Don't attempt to use just "any" solvent to clean disassembled carburetor parts. Buy a dedicated carburetor cleaner. These cleaners are available in spray cans and 1-gallon cans to 5-gallon buckets. Dedicated carburetor cleaner solvents are designed to remove varnish and other deposits.

To remove these deposits properly from exterior surfaces as well as small orifices and passages, it's best to soak the parts in this cleaning solution. The length of time varies, depending on the strength of the solution and the amount of deposit

buildup. Parts may require soaking anywhere from 10 minutes to as long as an hour. Try to avoid soaking for extended periods, such as overnight.

These solvents are extremely aggressive. Extended soaking can, depending on the specific solvent and the soak time, potentially oxidize the finish. If the surfaces appear white and "fuzzy" after soaking, try rinsing in undiluted vinegar, which may remove the surface oxidation residue.

If your cleaning basket isn't large enough to accommodate all of the parts, either use two baskets or soak half of the parts for the recommended time and soak the remaining parts during a second soak phase. Do not "help" the cleaning process by using a stiff wire brush, which can scratch precision surfaces. Allow the chemical to do its job.

Do not soak an assembled carburetor. Once the carburetor has been completely disassembled, load all parts into a basket (to avoid losing small parts) and soak as a group.

Do not soak rubber or elastomer items such as O-rings or diaphragms; the cleaning solution softens and destroys these materials.

Also avoid soaking electrical parts, such as heated choke assemblies or solenoids. Some late-model Holley carburetors have Teflon-coated throttle shafts. Expose these shafts in cleaning solution only long enough to remove deposits. Extended exposure to strong solvents can degrade the Teflon coating.

You need a tank or bucket large enough to accommodate your parts basket. A 3- or 5-gallon bucket of carburetor cleaner allows the basket to be dunked directly into the container. Another option is to use a parts-cleaner tank, filled with a dedicated carburetor cleaning solvent. Don't just "rinse" the parts in this solvent. Fully immerse the parts (in a steel parts basket) and allow them to soak.

When the parts have soaked long enough to remove all varnish and carbon deposits, rinse the parts in hot water and blow-dry with compressed air. My personal preference is to rinse in a mixture of hot water and Dawn dishwashing liquid, which does a great job of removing any residual solvents and oils. Follow that with a thorough hot-water rinse to remove the soap residue. Always blow all parts dry with compressed air.

Don't Use Air Tools

Never use any pneumatic (air) tools for carburetor disassembly or assembly. Carburetors are precision assemblies that should be serviced only with hand tools. The only application for compressed air involves cleaning; you can use a rubber-tipped blowgun as part of the process. An alternative to shop-type compressed air is to use a hand-held "dust remover" spray can of compressed air or compressed nitrogen. These spray cans are commonly available at office supply stores; they're intended for cleaning computer keyboards, camera lenses, etc. ■

When handling carburetor cleaner solvents, always wear protective rubber gloves and safety goggles to avoid skin and eye contact. Always keep the solvent away from any sparks or source of fire.

When you need to measure or verify various dimensions, such as baseplate thickness, linkage rod diameter, throttle bore diameter, etc., a quality dial caliper, micrometer, and small-bore gauges are necessary. If you plan to use drill bits to check, verify, or enlarge main jets, you must measure the drill bit shank with a micrometer to verify the actual diameter of the drill bit.

Media Blasting

Avoid media-blasting any carburetor part if at all possible. If it's necessary to media blast your carb, take all precautions to protect the carburetor. Improper media blasting as a cleaning method for carburetor components is downright dangerous because it can damage surfaces as well as potentially contaminate the carb.

In some instances, with proper care, certain types of media may be used to clean stubborn deposits. Soda blasting is the safest because soda is relatively benign (not too abrasive) and is water soluble for cleanup afterward. Items such as base plates or empty carburetor bowls and main bodies may be cleaned with walnut shell or corncob media, as long as you don't dwell the blast in concentrated areas.

Never sandblast any carburetor part; sand is far too abrasive. Aluminum oxide (in a glass bead cabinet, for example) may be used, but again, avoid concentrating the blast and be sure to use fairly low air pressure.

Avoid using any abrasive media for cleaning parts that have captive shafts because the media grit can contaminate the shafts and it can be difficult (or impossible) to flush out. Aggressive blasting can ruin precision orifices and mating surfaces and can prevent gaskets from sealing properly. Base assemblies and throttle shaft pivot points are easily contaminated; this can result in parts sticking and premature wear for shaft and shaft bore surfaces.

If you end up stripping fuel bowls and/or the main body down to a bare surface (by choice or by accident), the components may be re-plated with dichromate or zinc. A local plating shop can handle this, or the parts can be sent to Holley's restoration shop. For that matter, Holley offers complete and extremely high-quality rebuilding/refinishing services if you prefer not to perform a rebuild on your own.

Measuring Tools

Although not mandatory for most carburetor service, a selection of precision measuring tools can be handy for measuring and comparing various components. These include instruments such as dial or digital calipers, micrometers, and small-bore gauges that can measure baseplate thickness, throttle bore diameters, and so on. If you plan to enlarge any jets or other orifices with a drill bit, you must measure the bit first with a micrometer to verify its diameter.

Carburetor Case

Form-fitted hard-plastic carburetor cases are used primarily by racers who need to store and transport spare carburetors. They are also available for anyone who needs to safely store carburetors, prevent damage of valuable carburetors during transport, and to protect the carburetor from contamination.

Holley offers a black ABS case (PN 36-176). ■

Carburetor storage cases are extremely useful if you store or transport multiple carbs. A case protects the carb from damage as well as from dust and moisture contamination. This is Holley carb case (PN 36-176).

CARBURETOR DISASSEMBLY

If you plan to disassemble a carburetor, it should be done with the carburetor removed from the intake manifold and placed on a clean workbench. To make the carb more stable, place it on a stand that supports the carb by its baseplate-to-manifold mounting holes. This also makes the disassembly job much easier.

Various stand designs are available, including individual plastic stanchions that snap into the baseplate holes, cast-aluminum one-piece stands with four dowels that engage the baseplate holes, or vise-mounted stands that engage the baseplate mounting holes and allow the carb to pivot during service for easy access.

Allow the carb to cool completely before removing it from the intake manifold. Before removing the fuel inlet hose or line, place a small drain cup under the fitting(s) to catch any fuel that spills during removal of the fuel feed.

When the carb is removed from the engine, but before servicing, drain as much fuel as possible by rotating the carb slowly (sideways and upside down) over a fuel-safe drain pan. Some fuel will likely remain, but this eliminates most of the fuel present in the bowls and circuits.

Fuel Inlet Removal

On a 4150 or 4500 carburetor, the primary and secondary fuel bowl inlet fittings thread directly into the bowls; they use a male hex drive. Depending on the specific carburetor, either a 1- or 3/4-inch wrench is required. Models requiring a 1-inch wrench have little access between the hex and bowl. It's common to use a 1-inch open-end wrench, engaging and unscrewing in small increments because of the tight space. Carbs with the smaller 3/4-inch hex have more clearance for an open-end, box, or socket wrench.

AED's adjustable carb fixture adapts to any carb and allows securing to a bench vise. The carb can be rotated for easy access to the entire carb.

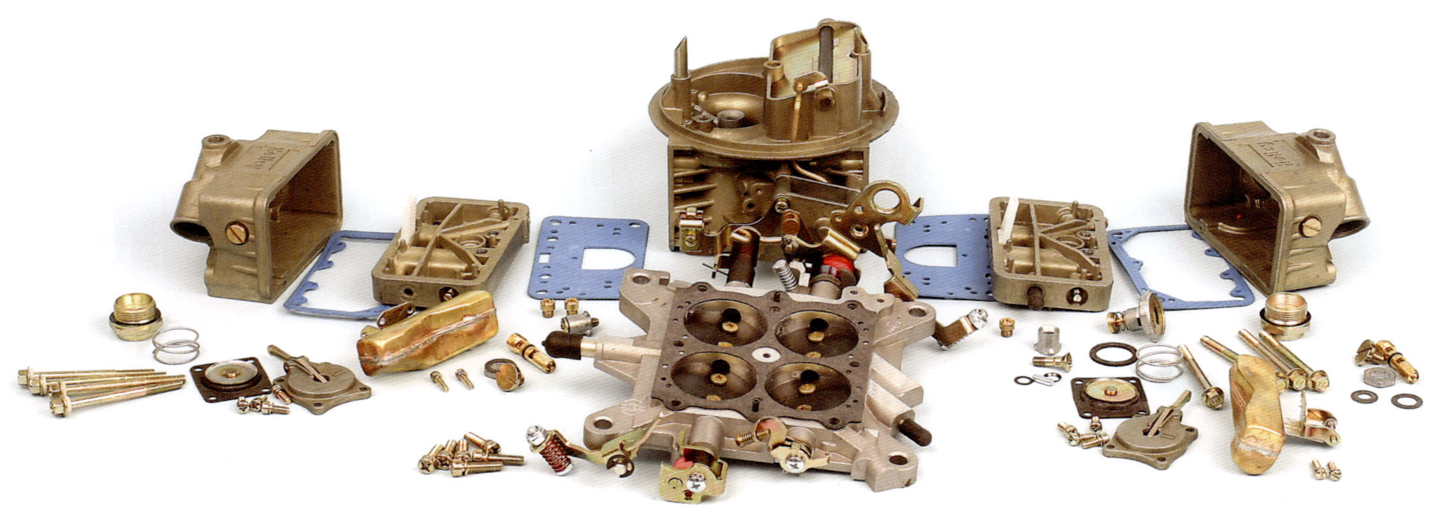

This 4150 Double Pumper carburetor is fully disassembled, except for the shafts and throttle plates from the baseplate. If you do not perform carburetor service on a regular basis, you can see why it's important to keep everything organized during disassembly. It's not a bad idea to take a few digital photos during your disassembly, and to bag and tag parts (primarily the float assembly, etc.) to avoid confusion during cleaning and reassembly.

The fuel inlet fittings have right-hand threads; rotate the fittings counterclockwise during removal. Depending on the specific carburetor, an internal in-line screen or sintered bronze fuel filter may be present. Be careful when removing the fittings so that you do not drop or lose a filter.

On 4160 carbs with a single fuel feed at the primary fuel bowl, a banjo-type fitting may be present on the driver's side of the fuel bowl. Using an 11/16-inch wrench, rotate the banjo bolt counterclockwise to loosen and then remove it.

Prior to disassembly, mount the carburetor on a working stand. This provides a stable platform on the workbench and provides clearance for the linkage.

Several types of carburetor stands are available. Inexpensive plastic stanchions such as this one simply snap into the baseplate mounting holes.

Each side of the banjo fitting includes a thin sealing washer. Be careful not to drop or lose these two sealing washers. The hex flats on the banjo bolt are very shallow. Use care when engaging the wrench; make sure that the flats are fully engaged to avoid damaging the hex head.

Fuel Inlet Removal

1   Remove Fuel Inlet Fittings

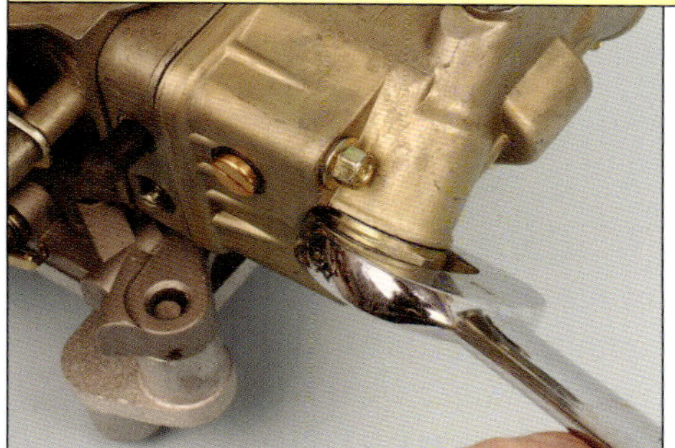

Remove the fuel inlet fittings before removing the fuel bowls. This allows you to hold the carb steady while cracking the fittings loose. Some 4150- and 4500-series fuel inlet fittings have a 1-inch hex. Clearance between the hex and the fuel bowl is tight, so you need an open-end wrench, not a common 1-inch socket or box wrench.

When removing the fuel inlet fitting, note that a thin metal crush washer is in place. This washer may stay with the fitting when removed, or it may stick to the fuel bowl inlet fitting port. If you don't see the washer on the fitting, be sure to retrieve it from the fuel bowl.

2   Remove Fuel Inlet Fittings *CONTINUED*

A 4160 car with a single fuel inlet on the primary bowl has a banjo-style fitting assembly. Remove the banjo bolt to free the assembly from the fuel bowl.

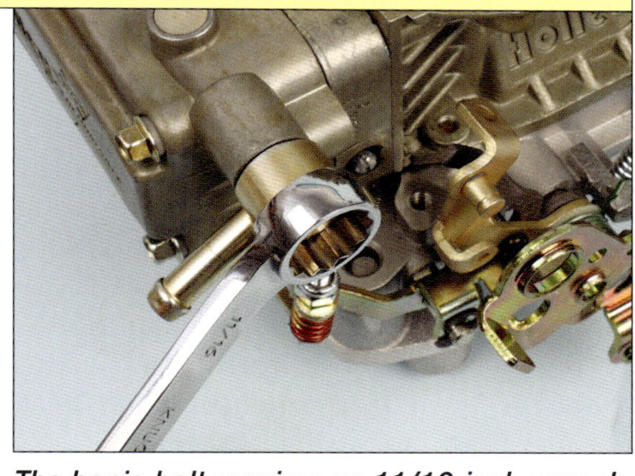

The banjo bolt requires an 11/16-inch wrench. A socket, box, or open-end wrench suffices.

3   Fully Engage Wrench

The banjo bolt's hex head surfaces are fairly thin. Make sure that the wrench is fully engaged to avoid burring the brass head.

4 Separate Fuel Fitting from Body

After you unthread the banjo bolt from the fuel bowl, remove the assembly, which includes the banjo bolt, banjo fitting, and filter. A thin crush washer is present on each side of the banjo fitting, one between the banjo bolt head and fitting and one between the banjo fitting and the fuel bowl.

5 Remove Fuel Inlet Fitting

Some 4150 carbs have fuel inlet fittings with a smaller 3/4-inch hex drive, requiring a 3/4-inch wrench. This smaller hex provides easier access than the 1-inch hex fitting used on other carbs. The smaller hex protrudes a bit farther for easier wrench engagement.

Fuel Bowl Removal

With the carb mounted on a support stand, place a shop towel under the primary fuel bowl to catch any residual fuel. Remove each of the four fuel-bowl screws, along with each screw's sealing washer using a 5/16-inch socket with a 1/4-inch-drive ratchet. This combination provides maximum clearance between the socket wrench and the fuel bowl. The washers may remain on the screws or they may stay stick to the fuel bowl. If a washer sticks to the fuel bowl, use a small flat-blade screwdriver to gently pop the washer free, and then place the washer(s) onto their respective fuel-bowl screws. This helps to keep everything in order and prevent losing any washers.

If you perform a disassembly for cleaning or rebuilding purposes, plan to use new gaskets (fuel bowl to metering block and block to main body). Even so, it's best to try to avoid damaging the original gaskets. You may be forced to reuse one in a pinch, and they are great for identifying and matching to the new gaskets.

Gently pry each gasket loose. Usually, with enough patience and care, you can remove them without tearing. Set the gaskets aside, organizing them next to the fuel bowl and metering block.

Carb Orientation

During carb disassembly, try to keep the carb oriented in one direction. For example, keep the primary side on your left and the secondary side on your right. If you're new to carburetor work, this helps you to keep parts organized and identifiable regarding their location on the carb. ■

1 Remove Float Bowl Screws

Years ago, some Holley fuel-bowl screws had a slotted head that required using a flat-blade screwdriver. The common screws used today have a hex head, which require a 5/16-inch wrench. Crack each screw loose, and then continue to remove all four. Each fuel-bowl screw seals to the fuel bowl with a sealing washer. If the washer does not stay with the screw, it's likely stuck onto the fuel bowl. Make sure to retrieve all washers and keep them with the screws.

2 | Separate Fuel Bowl from Main Body

Professional Mechanic Tip

The gasket may cause the fuel bowl to stick. Do not jam a flat-blade screwdriver or chisel between the fuel bowl and metering block, or between the metering block and main body. This can result in gouges at the mating edges, which potentially create raised burrs that prevent proper sealing during reassembly. A light tap with a plastic hammer pops it loose. Avoid digging between the fuel bowl and metering block with a screwdriver or chisel, as this can gouge the mating edges.

3 | Remove Fuel Bowl Gasket

If you want to save the original gasket, carefully peel it off the metering block. Several small dowel pins locate the gasket. Peeling the gasket off in a hurry can easily result in tearing the gasket at the dowel locations.

4 | Inspect Metering Block

Critical Inspection

When you have a metering block (primary side on all carbs, and on the rear of 4150 and 4500 carbs), no additional screws are present. With the fuel bowl removed, only the gasket between the metering block and main body hold the metering block in place. Set the primary fuel bowl aside.

5 | Remove Metering Block

A few taps with a plastic hammer should easily dislodge the metering block from the main body. Quite often, simply wiggling and pulling with your fingers is enough to remove the block.

6 | Remove Metering Block Gasket

Use care when removing the gasket that sealed the metering block to the body. You should be able to remove the gasket intact if you're careful.

7 Inspect Main Body Casting

The primary fuel bowl, gaskets, and metering block have been removed, so at this stage, inspect the main body casting for any damage or unusual wear.

8 Remove Fuel Bowl Sight Plugs

Remove the brass fuel bowl sight plugs by rotating each plug counterclockwise. Make sure that the screwdriver is seated fully and squarely to avoid burring the screw's drive slot. A thin crush washer seals the plug. Save the washer in case you decide to reuse it. Keep the sight-bowl screw and its washer bagged together to avoid losing. If the bowl has a glass sight window (as found on Ultra and HP models for example), a C-clip secures the glass. Squeeze the clip using C-clip pliers and remove the glass and the seal.

Metering Bowl Disassembly

If your carburetor is a 4150 or 4500, removal of the fuel bowl and metering block is identical to the steps you used during the primary-side service.

If your carb is a 4160, you may have an external fuel transfer tube on the driver's side. With the primary fuel bowl removed, you can simply pull the tube out of the secondary fuel bowl. The sealing gaskets at each end of the tube remain in the fuel bowls. Using a small flat-blade screwdriver, gently pry these rubber seals out of the front and rear bowls and set aside.

The 4160 carb primary fuel bowl is secured to the main body with four 12-24 x 2½–inch screws, just as with the primary and secondary bowl screws found on 4150 and 4500 carbs.

However, because the secondary side of a 4160 carb has a thin metering plate instead of a metering block, the four 12-24 secondary bowl screws are shorter (1⅞ inches) than those for the primary fuel bowl. Simply be aware of this. The longer 2½-inch bowl screws that are required for bowls with metering blocks cannot be used at the secondary side of a 4160 carb with a metering plate.

The 4160 carburetors have a metering plate at the secondary side, which is secured independently to the main body with six flat-top clutch-head machine screws, size 8-32 x 1/2 inch. Although a small flat-blade screwdriver can be used to service these screws, the proper choice is a 5/32-inch clutch-head driver, which provides a positive engagement with the screw and prevents possible slipping and gouging.

The metering plate is likely to be slightly stuck to its gasket. Gently pry on the plate to remove it. Do not try to jam a flat-blade screwdriver between the plate and main body because it may cause scratches, gouges, or burrs that can prevent proper sealing during reassembly. Again, try to keep all parts organized and in order relative to the primary and secondary sides.

1 Inspect Metering Plate

A 4160 carb has a metering plate on the secondary side, instead of a jet-equipped metering block.

2 | Remove Metering Plate Screws

The metering plate is secured to the main body with six clutch-head screws.

Although a small screwdriver may be used to remove clutch-head screws, the correct tool is a 5/32-inch clutch-head driver. This positively engages the hourglass-shape screw drive, avoiding the potential for damaging the screw.

Critical Inspection

3 | Inspect Fuel Transfer Tube

This 4160 has a single fuel feed at the primary bowl and an external fuel transfer tube that routes fuel to the secondary bowl from the primary bowl. Each end of this tube seats into a rubber grommet seal. When removing one of the fuel bowls, the tube sticks to one bowl and leaves the opposite bowl.

4 | Remove Fuel Transfer Tube

Gently twist and pull each tube free from its bowl. The tube exits easily.

5 | Remove Sealing Grommet from Fuel Bowl

Use a pick or small screwdriver to pull the sealing grommet from each fuel bowl. You can keep the seals as a reference, but you should always install new seals during assembly.

Float Removal

The 4150 and 4500 carbs have center-hung floats; the 4160 carbs have side-hung floats.

To remove a center-hung float, remove the two 6-32 x 1/2–inch screws that secure the float hinge to the bowl. This requires a small flat-blade screwdriver.

A side-hung float pivots on a horizontal pin. A light spring on the underside of the float arm provides a support assist. To remove the float, carefully remove the small C-clip that secures the float arm to the horizontal pin. Use a very small flat-blade screwdriver or a pick to pull the C-clip out of its groove on the stationary pin (a magnetic-tip screwdriver is a good idea to prevent dropping the clip). It's a good idea to store the float, spring, and C-clip in a Ziploc bag.

The needle and seat assembly is housed in a white nylon block, with a small wire clip that engages into the float's upper arm. Once the float is removed, the needle and seat easily fall out of the bore.

1 Remove Fuel Bowl Float

A center-hung fuel bowl, as on 4150- and 4500-series carbs, secures to the inside upper center of the fuel bowl with two screws. Use a small flat-blade screwdriver to remove the two 6-32 x 1/2–inch float hinge screws. After unthreading the screws, if you have difficulty retrieving them while reaching in with your fingers, use needle-nose pliers, a pencil magnet, or a hemostat.

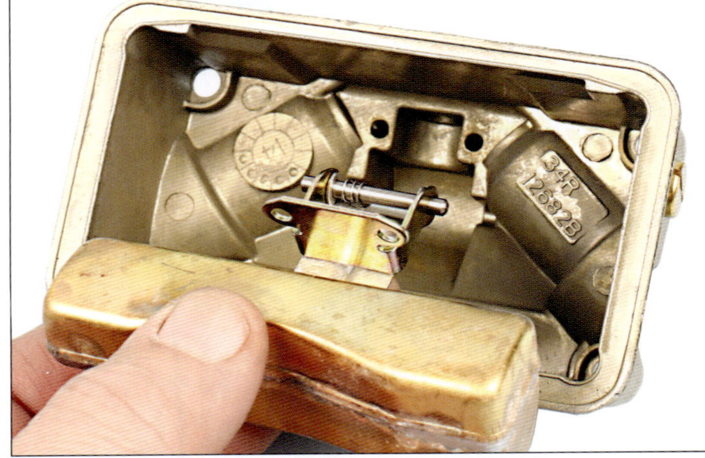

A lightly spring-loaded hinge is attached to the float. Remove the float and hinge assembly as a unit. Keep the float and mounting screws together in a Ziploc bag to avoid losing the screws.

2 Inspect Float

A side-hung float, as found on 4160 carbs, pivots on a pin located to one side, near the needle and seat. This is a secondary bowl from a 4160 carb.

3 Remove C-Clip on Float Armature

The pivot pin for the float hinge has a small C-clip. This clip must be removed to remove the side-hung float.

4 Remove C-Clip on Float Armature CONTINUED

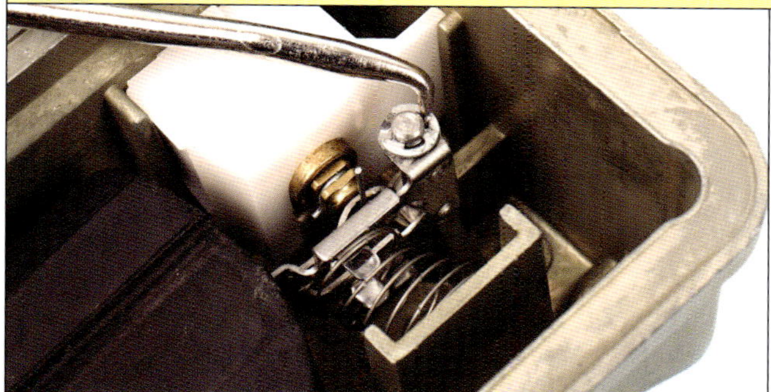

Use a small flat-blade screwdriver or a pick to dislodge the small C-clip. Be careful not to exert too much force, to avoid popping the clip off abruptly. It's small and easy to lose.

5 Remove C-Clip on Float Armature CONTINUED

Use a small pencil magnet or magnetic-tipped screwdriver to retrieve the clip without dropping it. It's a good idea to place the float and C-clip into a small plastic bag to avoid losing the clip.

6 Remove Side-Hung Float from Fuel Bowl

On a side-hung float, the light-assist spring is attached to the float hinge arm.

Needle and Seat Removal (4150)

On a 4150 or 4500 carb, the needle and seat assembly in the primary and secondary fuel bowls are located at the top center of each fuel bowl.

1 Remove Float Bowl Needle and Seat

On center-hung float bowls, access the needle and seat assembly at the top exterior of the bowl. The screw serves as a lock to hold the adjusted position of the needle and seat. The hex provides needle and seat height adjustment. While holding the hex nut stationary with a 5/8-inch wrench, rotate the screw counterclockwise with a flat-blade screwdriver to loosen the hex.

2 Remove Float Bowl Screw and Hex Nut

Remove the locking screw and hex nut. Remove the needle and seat assembly by rotating it counterclockwise. (The needle and seat upper section threads into the fuel bowl.)

3 Remove Needle and Seat from Port

Turn the needle and seat assembly counter-clockwise until the threads have cleared the threaded section of the casting. Once it's unthreaded from the fuel bowl, pull the needle and seat assembly out of its port.

4 Inspect Adjuster Hex Nut

The adjuster hex nut has two opposing flats in the threaded hole. This engages onto the flats of the needle and seat assembly, allowing you to adjust the needle and seat up or down.

Instead of using a flat-blade screwdriver and a 5/8-inch wrench, this specialty needle and seat tool from Willy's Carburetors makes the job easier and quicker. The tool's female hex opening engages the needle and seat hex nut, while a straight blade on the tool's inner shaft engages the locking screw. Hold the tool's knurled aluminum body stationary and turn the black knob at the top of the tool to unthread the locking screw. Then turn the knurled body of the tool to raise and remove the needle and seat assembly.

The slot-driver shaft and outer tool body are independent, allowing you to perform both tasks of screw loosening and nut removal with one tool. After you unthread the needle and seat, remove the entire assembly, including locking screw, adjuster nut, and needle and seat.

Needle and Seat Removal (4160)

On a 4160 carb, the needle and seat must be serviced with the fuel bowl removed. With the side-hung float removed, the needle and seat assembly is easily removed from its vertical overhead housing.

1 Remove Needle and Seat Assembly

After the float has been removed, access the needle and seat assembly on a side-hung float bowl from inside the bowl. Inside a fuel-safe plastic housing, the needle and seat assembly has a wire clip that engages the float's hinge arm.

2 Remove Needle and Seat Assembly CONTINUED

Simply turn the fuel bowl upright so the needle assembly can fall out of its housing bore.

Accelerator Pump Removal

The accelerator pump on the primary fuel bowl (and on the secondary fuel bowl of a Double Pumper carb) is secured with four 8-32 screws. The screws' length on 4150 and 4160 carbs is 3/8 inch; 4500 carbs tend to use a thicker housing, requiring a screw length of 9/16 inch.

A plastic cam is located at the throttle shaft where the accelerator pump operating lever makes contact. The shape of the cam dictates how quickly and how long the accelerator pump operates. It is held in place by a single screw that is removed with a flat-blade screwdriver.

1 Remove Accelerator Pump

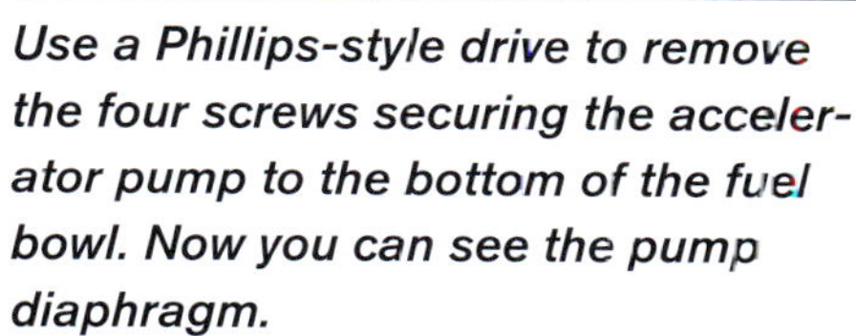

Use a Phillips-style drive to remove the four screws securing the accelerator pump to the bottom of the fuel bowl. Now you can see the pump diaphragm.

Professional Mechanic Tip

2 Remove Accelerator Pump Diaphragm

If you plan to reuse the original diaphragm, carefully peel the rubber from the fuel bowl.

3 Lift Diaphragm Spring from Casting

With the pump diaphragm removed, you see a light-assist spring. Simply remove the spring with your fingers.

4 Remove Check Valve from Casting

The orange "dot" is a check valve. When the accelerator pump arm presses on the diaphragm, this silicone/rubber check valve is pushed upward, applying pressure to the fuel in the bowl, sending the extra shot of fuel up to and through the pump's discharge squirter nozzle. If it's not leaking, and if you don't plan to immerse the bowl in carb cleaner, there's no reason to disturb this. This valve is secured to the bowl with a formed peg that passes through the floor of the bowl. If you pry it out, chances are high that you'll damage it, so plan to replace it if you decide to remove it.

Critical Inspection

5 Inspect Check Valve

The silicone/rubber check valve has a small nipple that engages the valve to the fuel bowl. Note the small orange head.

6 Remove Accelerator Pump Check Valve

Simply pry the face up with a fingernail and pull/wiggle it loose. If you tear the valve's nipple in the process, replace it with a new valve from a Holley rebuild kit. The check valve's nipple has a radiused barb tip that secures the valve in the bowl.

Critical Inspection

7 Inspect Accelerator Pump Cam

The accelerator pump plastic cam is located on the throttle shaft. These typically remain in good condition, but check to make sure there is no damage that would inhibit operation.

8 Inspect Throttle Lever

The accelerator pump's throttle lever rides on the cam, influencing how quickly the pump reacts as well as pump duration. You need to verify that it operates as designed.

9 Remove Accelerator Pump Cam Screw

The accelerator pump cam is secured to the main throttle lever with a single flat-top screw.

Using a small flat-blade screwdriver, remove the screw and wiggle the cam from the throttle shaft.

10 Select Accelerator Pump Cam

Accelerator pump cams are available in different profiles for accelerator pump tuning purposes. Multiple screw mounting holes are also provided to further tune the cam position (similar in concept to advancing or retarding an engine's camshaft).

11 Inspect Accelerator Pump Throttle C-Clip

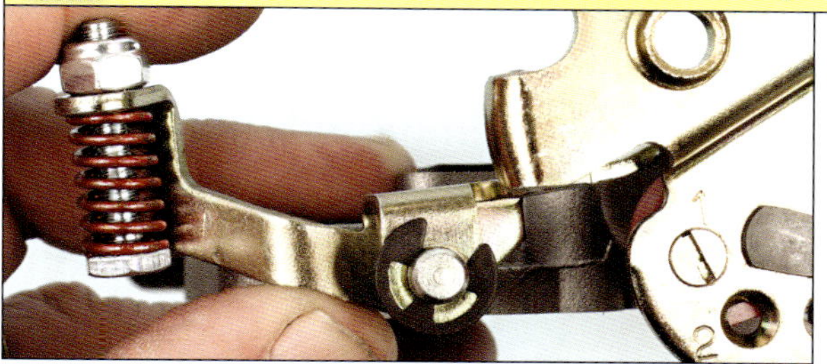

The accelerator pump throttle lever is held in place on the pivot pin with a steel C-clip. The pivot pin has a groove that accepts the clip.

12 Remove Accelerator Pump Throttle C-Clip

Pry the C-clip away from the pivot pin with a small flat-blade screwdriver. Be careful not to pop it off too quickly, because it's easy to lose. Push it off about halfway, and then grab it with needle-nose pliers or a hemostat to avoid dropping it.

13 Remove Accelerator Pump Lever

Once the C-clip is removed, slide the lever off its pivot pin.

14 Inspect Accelerator Pump Throttle Lever Pivot

The accelerator pump throttle lever arm pivots on this pin. If the pin is worn and must be replaced, use a 5/16-inch wrench and rotate counterclockwise to remove it.

Accelerator Pump Discharge Nozzle Removal

The discharge nozzle, on either the primary or secondary side, is secured with a single 12-28 Phillips-head screw at the top of the main body between the front two barrels (and between the rear two barrels on a Double Pumper). Hold the discharge squirter steady and remove the screw. One small, thin metal sealing washer is under the screw head and one is at the bottom of the nozzle unit. It's very easy to overlook the lower washer because it can be left behind or dropped when the squirter is removed.

Under the discharge nozzle, inside the fuel passage, is a small pump discharge needle valve. Turn the main body upside-down to let this needle valve drop out. If you're not aware of the presence of this small needle valve, it may fall out later when you turn the main body upside down; you'll either not notice it at all or wonder where it came from. It's a good idea to place the squirter, screw, both washers, and the needle valve in a small Ziploc bag to keep them together and prevent losing anything.

1 Remove Accelerator Pump Discharge Nozzle

Use a Phillips screwdriver to remove the single machine screw that secures the accelerator pump's squirter discharge nozzle to the main body. This nozzle is being removed from the secondary side of a 4150 carb.

Professional Mechanic Tip 

2 Remove Accelerator Nozzle Gasket

The discharge nozzle has two small, thin metal gaskets, one at the top between the screw head and nozzle and one at the base between the nozzle and main body. It's easy to miss the lower gasket, which often remains on the main body. You can easily retrieve it using a small pencil magnet. If you drop it into the venturi, simply open the throttle plate and allow it to drop onto the workbench.

3 Locate Discharge Nozzle

Access to the discharge nozzle on the primary side of a carb with a choke is somewhat limited because of the choke plate. Be patient and use a screwdriver to unthread the discharge nozzle screw.

4 Remove Discharge Nozzle and Screw

Use needle-nose pliers or a hemostat to retrieve the nozzle and screw as an assembly. Use a pencil magnet to grab the nozzle's base gasket.

5 Remove Discharge Nozzle Needle Valve

Remove the small needle valve inside the nozzle's screw hole passage. The easiest method is by carefully turning the carb upside-down onto a clean rag on the workbench. If you do not remove it now, you may lose it the next time the carb is turned upside-down.

6 Organize Discharge Nozzle Parts

To keep things organized, bag the discharge nozzle, screw, gaskets, and needle valve together.

Jet Removal

Main metering jets have a straight driver slot. A flat-blade screwdriver may be used, but a much better tool choice is a dedicated jet driver such as the one offered by Holley. This tool has a male tang that engages the slot and a centering tip that keeps the tool centered on the jet, which eliminates the chance of the tool slipping off the slot during removal or installation. Engage the tool and rotate it counterclockwise to remove the jets.

1 Select Main Jet Wrench

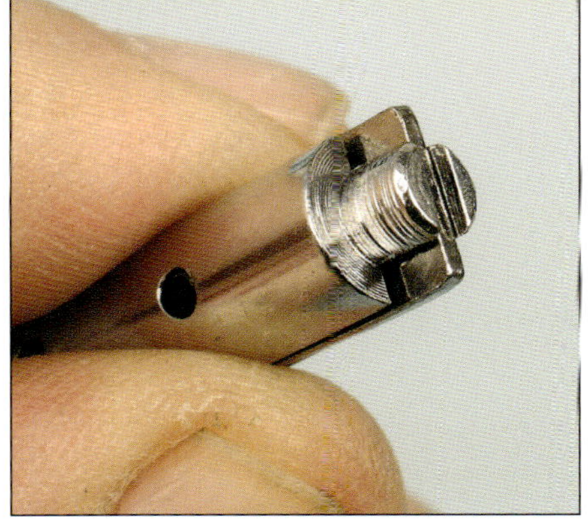

Although a flat-blade screwdriver can be used to remove or install main jets, a much better choice is a dedicated jet driver such as those offered by Holley and a few other tool makers.

2 Remove Main Jet

This tool has a center round dowel that automatically centers the tool into the face opening of the jet; the flat driver seats into the jet slot. The tool prevents slipping, which can occur with a common flat-blade screwdriver. The tool's center dowel keeps the driver securely positioned.

Power Valve Removal

The power valve is threaded into the metering block from the rear. The body of the power valve has a four-sided flat drive instead of a hex. A 1-inch wrench is required for removal or installation. The flats on the power valve are fairly shallow. Although a conventional open-end, box, or socket wrench may be used, a superior choice is a dedicated power-valve wrench. This is a billet aluminum tool that is specifically designed to perfectly engage the power valve without the danger of slipping off or gouging the flats or nearby surfaces.

Rotate the power valve counterclockwise to remove it. The power valve includes a sealing gasket. If you plan to reuse the existing power valve, make note of the gasket style (open center or center with three tangs) because you must use the same style of gasket for the power valve.

1 Remove Power Valve

Power valves have a four-sided flat, requiring a 1-inch wrench. However, the drive flats are fairly shallow, so care must be taken to make sure that the wrench remains fully seated. A dedicated power-valve drive tool, such as this one from Willy's Carburetors, eases the job. The billet aluminum power-valve tool is designed specifically for servicing power valves. Insert the tool onto the power valve, holding the tool flush with the metering block and rotate the tool counterclockwise. The knurled grip provides excellent finger traction.

2 Remove Power Valve CONTINUED

As this photo illustrates, the power-valve tool is not a common 1-inch wrench and is designed specifically to securely engage a Holley power valve. Because the tool is aluminum, it doesn't gouge the metering block.

If you plan to perform carb service on a regular basis, this is a must-have tool.

Overflow Whistle Removal

The white plastic overflow whistle is located on the upper area of the metering block (if so equipped) and is held in its metering block slot with a single interference-fit pin.

1 Remove Overflow Whistle Pin

To remove the pin that secures the plastic overflow whistle, insert a flat-blade screwdriver into the rear of the cavity and pry upward to push the screw up. The self-tapping screw is tiny, so bag it for reference. Once the screw has been pushed upward, you may be able to wiggle it free with your fingers. If not, use needle-nose pliers or a hemostat.

2 Remove Overflow Whistle

After the locating pin has been removed, you can easily pull the overflow whistle from the metering block.

Mixture Screw Removal from Metering Block

The needle screw is sealed with a small round gasket in the metering block. You can remove it using these two simple steps.

Professional Mechanic Tip

1 Remove Idle-Mixture Screw

Rotate the idle-mixture screw counterclockwise for removal. If you plan to rebuild the carb, or if you've had a leak at the screw, remove the gasket with a pick.

2 Remove Idle-Mixture Screw Gasket

If you plan to clean the metering block with carb solvent, remove the small sealing gasket and replace it with a new gasket.

Choke Assembly Removal

For carburetors that are equipped with an electric choke, the black-plastic electric choke housing is secured with three 8-32 x 3/8–inch Phillips-head screws. A tension washer maintains pressure on the housing to prevent the housing from accidentally rotating.

An eyelet on the flat-wound spring inside the choke housing engages the choke plate lever. Pull the choke housing straight off; avoid cocking it during removal to prevent potential spring damage.

The choke housing's cast-metal base is attached to the main body with three 8-32 x 1⅛–inch Phillips screws. Before removing these screws, remove the small hairpin that secures the vertical choke rod to the choke lever assembly. Use small needle-nose pliers or a hemostat to remove this pin.

When removing the base, note the small orifice with a small round gasket on the backside of the base. This influences manifold vacuum, which circulates air through the housing. In turn, the pull-off piston activates inside the base and aids in opening the choke plate after the engine warms up. Inspect this small gasket. If damaged, it may be the cause of a formerly "mysterious" vacuum leak.

Electric Choke Dissassembly

An electric choke has a black plastic thermal-spring housing that's secured to the choke lever assembly with three screws. A small hairpin secures the choke rod to the choke assembly lever.

1 Remove Electric Choke

Remove the three screws and the black spring housing along with the spring-steel three-hole washer. The outer tension washer applies pressure to maintain the choke housing's clock position once it's adjusted.

2 Remove Choke Gasket

The large flat gasket should remove easily, unless heat and age have caused it to stick. If so, use a razor blade to scrape the gasket free.

3 Inspect Choke Spring

A tang connects to the choke spring eyelet (upper right inside the casting). This spring inside the choke housing is in its rest position. It pushed the choke plate rod up and caused the choke plate to close. As the spring heats, the lever pulls the choke rod down and opens the choke plate.

Professional Mechanic Tip

4 Remove Cotter Pin from Choke Rod

Before removing the electric choke's cast base from the main body, remove the small cotter pin that secures the choke rod to the choke assembly lever. Needle-nose pliers or a hemostat makes this easy. Using a hemostat allows you to not only grab the pin, but because a hemostat locks, the pin is securely retained during removal. A hemostat includes a serrated locking tang between the finger rings. To free the spring clip from its grip simply squeeze the tool together and dislocate its locking serrations.

Mechanical Choke Disassembly

A mechanical choke assembly attaches to the main body in the same manner as an electric choke. It's secured with three screws, and the choke plate rod engages to a lever on the choke assembly, which is pinned with a small hairpin. Remove the pin before removing the three mounting screws.

1 Separate Choke Assembly from Main Body

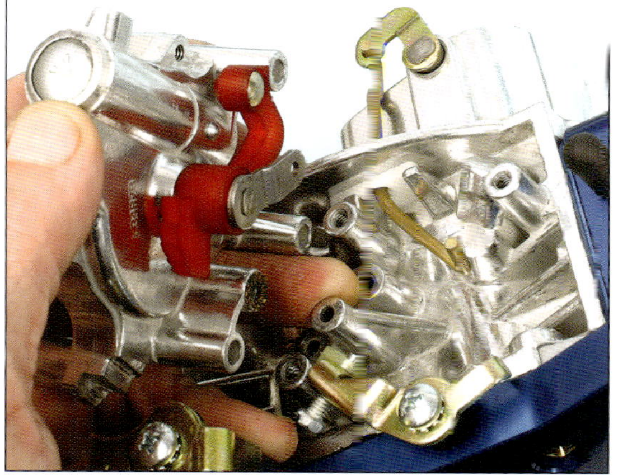

Remove the three screws holding the cast choke base assembly to the main body.

2 Inspect Air Vent Tubes

Critical Inspection

Note the air vent tubes on the base casting (at left). The larger tube at the lower left is the air "intake" tube that allows outside air into the housing. The smaller tube just inside the round base is a vacuum port that feeds from manifold vacuum to assist the pull-off piston that aids in opening the choke plate when the engine warms up.

3 Inspect Air Vent Tube Gasket

A small gasket seals the vacuum port that provides vacuum to the pull-off piston inside the choke housing base. A damaged gasket could explain a vacuum leak that you previously couldn't find. Regardless, before reinstalling the base, always install a new gasket in this location.

4 Remove Choke Armature from Main Body

Locate the three screws that mount the choke lever base on a 4150-series carb to the main body.

5 Remove Choke Rod

Before removing the three choke lever assembly screws, note that the choke plate-actuating rod engages to a lever on the bolt-on assembly. The choke rod is secured with a small hairpin that must be removed first.

6 Remove Choke Lever Assembly

Use a Phillips screwdriver to remove the three choke lever assembly screws from the main body.

7 Remove Choke Lever Assembly *CONTINUED*

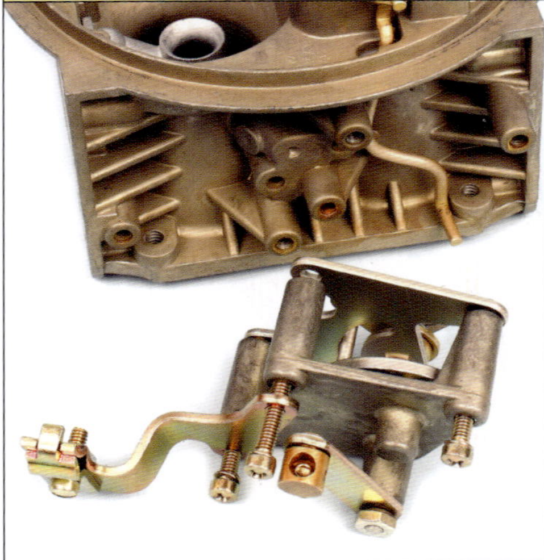

Remove the choke lever mechanism. The formed bracket at the rear (left side of photo) shares two of the mounting screws. This bracket includes a cable pinch mount for the manual-choke cable.

8 Inspect Choke Plate Rod

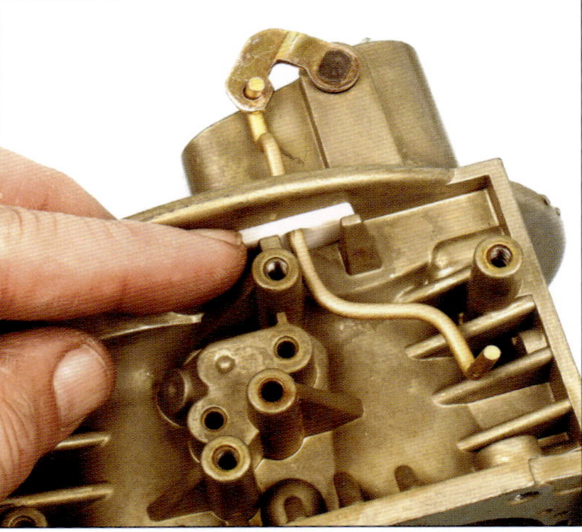

The choke plate rod has a nylon plate that helps to guide the rod with reduced friction.

Critical Inspection

9 Inspect Choke Plate for Damage

The choke plate rod may seem loose now, but it's captured at the top where the rod's 90-degree bend engages the choke plate lever. The rod can't be removed without bending the rod or the upper lever, which should be avoided. Of course, the choke plate could be removed, but that presents a problem because the choke plate is staked to the choke plate shaft. If there's no need to remove the choke plate (damage, etc.), there's no reason to remove this rod.

Vacuum Secondary Diaphragm Removal

The vacuum secondary assist unit mounts to the passenger's side of the main body, secured with three Phillips-head screws. The choke assembly base hides access to the forward-most screw, so the choke assembly must be removed first.

A single Phillips-head screw secures the bottom of the vacuum secondary actuator rod that connects to the secondary throttle plate shaft. Remove this screw first.

Remove the three screws that secure the vacuum assembly to the main body, and then wiggle the actuator rod's lower lever off of the secondary throttle shaft. This lever is engaged to the shaft via a two-sided flat design and the fit may be fairly tight. If necessary, use a small flat-blade screwdriver as a pry tool between the baseplate and lever. Pry gently, little by little, on opposite sides of the lever until it is free.

A small gasket seals the vacuum feed on the backside of the vacuum housing. On a "traditional" Holley, the lid of the vacuum housing is secured with four 8-32 x 1–inch Phillips-head screws. Remove these four screws and lift off the lid and spring. The diaphragm now pops up like an upside-down umbrella. This is a good example to justify upgrading to Holley's latest vacuum lid assembly (Holley part numbers 20-59 or 20-73); the spring can be changed without disturbing the diaphragm. This newer design is standard on 4150 Avenger carbs.

If the diaphragm needs to be replaced, the actuator rod is secured to the lower lever by a very small C-clip. Removing the C-clip and separating the lower lever from the rod allows you to lift the diaphragm out of the vacuum housing.

1 Locate Vacuum Secondary Screws

The vacuum secondary unit is secured to the main body with three screws and to the secondary throttle shaft with a single screw. Remove the choke assembly to gain access to the forward most of the three screws that hold the unit to the main body.

2 Remove Rod Lever Screw

Loosen the three mounting screws, but wait to remove the three housing mounting screws until you remove the lower screw that holds the rod lever to the throttle shaft.

3 Remove Vacuum Actuator Rod Screw

Remove the lower screw that secures the vacuum actuator rod lever to the secondary throttle shaft.

4 Remove Vacuum Secondary

Important!

The lever is positioned onto the throttle shaft with engaging flats. The lever may be a bit tight, so be patient and wiggle the lever loose from the shaft. Slight prying may be needed, but be careful not to bend the lever.

Remove the vacuum secondary unit from the carb.

5 Inspect Vacuum Port Gasket

The vacuum secondary assembly has an internal diaphragm that is an integral part of the actuator rod. Notice the small port with a gasket on the backside of the vacuum unit. This aligns with the vacuum port on the main body. Always plan to replace this gasket.

The vacuum port on the main body feeds the vacuum unit.

6 Remove Vacuum Assembly Top Cover

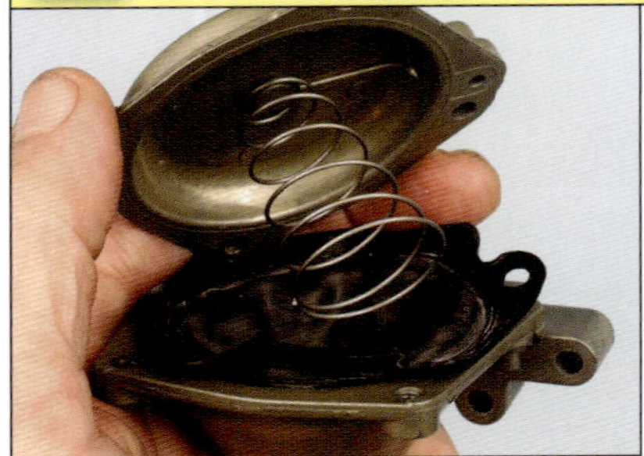

The vacuum unit's lid is secured with four 8-32 x 1–inch screws. Removing the vacuum housing's lid reveals the spring and the diaphragm.

7 Remove Diaphragm

The diaphragm and actuator rod is a pre-assembled unit. Remove the lower rod lever from the rod to remove the diaphragm from the housing.

8 Remove Actuator Rod C-Clip

The vacuum secondary actuator rod is secured to the throttle shaft lever with a very small C-clip.

9 Replace Spring

The Avenger series has a modular vacuum secondary assembly that allows the spring to be replaced without the need to remove the entire vacuum unit from the carb. In addition to serving as standard equipment on the Avenger series, this upgrade is available as a kit to convert any Holley vacuum secondary unit to this modular style.

Remove the two screws that hold the spring lid and the spring in place. The diaphragm remains contained under a separate cover. If you only want to change springs, this design avoids having to deal with disturbing and re-seating the diaphragm.

Main Body and Baseplate Separation

The main body is secured to the baseplate/throttle body with six 12-24 x 3/4–inch screws. These screws have a Phillips-style head. With all six screws removed, the main body should part easily from the throttle body.

As mentioned earlier, the accelerator pump levers and cams are easy to remove from the baseplate. If the throttle shafts appear to be sticking back or are worn, the throttle plates must be removed from the shafts so you can slide the shafts out of the baseplate. However, each throttle plate is secured to the shaft with a pair of small screws. The opposite side of the screw is deformed at the factory to prevent accidental loosening.

To remove these screws, use a die grinder to carefully grind off the protruding end of each screw so that it is flush with the shaft surface. Use extreme care to avoid damaging the shafts. If the throttle shafts are operating as designed, leave them alone and don't disturb the throttle-plate screws.

1 Remove Base Plate Screws

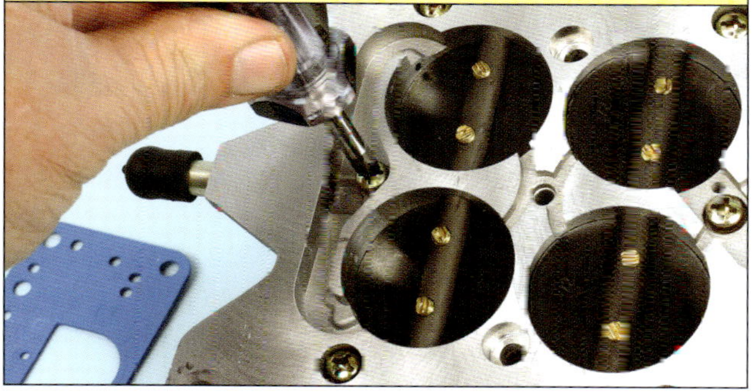

Use a Phillips-style driver to remove all six baseplate screws. Be careful to avoid slipping the driver to prevent gouging or scratching the baseplate's manifold mating surface. Each baseplate screw has a lock washer.

2 Inspect Baseplate

This is an overhead view of the baseplate (also called the throttle body) from a 4150 Double Pumper carb.

3 Inspect Main Body

This is the bottom view of the main body of a 4150 carb.

4 Remove Throttle Plate Screws

Each throttle plate is secured to the throttle shaft with a pair of screws. Grind off the opposite ends of the screws to remove them.

The screw tips are mashed/deformed to prevent the screws from loosening. If you must remove the throttle plates, use great care when grinding these down to avoid damaging the throttle shaft. Also, if you do remove the throttle plates, keep them organized for venturi bore location.

CLEANING AND INSPECTION

Carburetor cleanliness is absolutely critical. Any small particles such as dirt, sand, etc. can easily clog small ports and passages. Buildup of carbon, old grease, and so on can hinder movement of shafts and other moving parts. During any service or rebuild, take the time to ensure that every single component is absolutely clean and free of contaminants. Although a parts cleaner certainly comes in handy, it's highly recommended to use a dedicated solvent that is specifically designed for carburetor cleaning. You also need a few cleaning brushes of various sizes that should be dedicated to carb cleaning only to eliminate cross-contamination. Old general-use brushes may contain grit and other deposits because they may have been used to clean various engine, brake, suspension, or other parts.

You also need a source of compressed air to blow small passages clean. This can be accomplished with a shop air compressor or hand-held bottles of compressed air that are available at office supply stores (the same type used for cleaning computers).

Solvents

Although a spray-can of a brake-cleaning solvent may provide a reasonable cleaning choice, a specific carburetor cleaning solvent is a better choice because of its formulation. This cleaner, along with a small soft-bristle brush, can accomplish your goal. However, to remove even the most stubborn varnish or carbon buildups, a specially formulated carburetor cleaner, such as CRC Industries' TYME-1 provides superior cleaning performance. This solvent is available in both 1- and 5-gallon buckets.

TYME-1 is rather aggressive and is intended for metal parts only. Any soft parts, such as O-rings, gaskets, and diaphragms should be replaced. Even though some O-rings and diaphragms may appear reusable, if you've completely disassembled the

Load the carb parts in a parts cleaning basket. Although many small parts can be placed in the basket including jets, screws, sight plugs, and so on without fear of losing them, extremely small parts such as C-clips, pin clips, etc. are better cleaned by hand. Always wear chemical-resistant gloves and eye protection.

Lower the basket into the solvent bucket slowly to avoid splashing. Leave the basket submerged in the solvent bucket for about 20 minutes. After removing the basket, rinse all parts with hot water and blow-dry with compressed air. When blowing compressed air into the throttle body, remember to adjust your compressor to approximately 25 psi before blowing air into passages inside the main body.

When purchasing a can of cleaner, be sure to buy a solvent that is specifically intended for carburetor cleaning.

Be sure to squirt carb cleaner into each and every passage and orifice, and then blow everything clear with compressed air.

A superior cleaning solvent example is CRC Industries' TYME-1, available in 1- or 5-gallon buckets. A dedicated cleaning basket (shown) drops inside a 5-gallon bucket. The tall handle on the basket prevents the need to retrieve parts with your hands.

carburetor, it just makes sense to do it properly and start off with fresh seals, diaphragms, and gaskets.

Dedicated carb solvents, such as TYME-1, are very potent. You must wear protective gloves and safety glasses when dealing with this chemical. Just immersing the parts in a bucket of this solvent removes contaminants. Depending on the severity of the buildup, a soak time of 15 minutes to an hour usually does the trick. To avoid losing small parts in the bucket, and to prevent having to dip your hand into the solvent to retrieve parts, place all parts in a parts-washer cleaning basket. The basket is made of perforated steel that allows the solvent to drain when it's removed from the bucket; it also

Professional Cleaning Options

If you want to refinish the main body, metering block, and fuel bowls, especially for a factory-appearance restoration, Holley offers an in-house refinishing service that does outstanding work. A few years ago, I rebuilt an original 302 engine for a 1969 Camaro Z28. As part of the engine restoration, I sent the carb components to Holley for refinishing in the original "yellow zinc." They came back looking brand new and period correct.

Yes, aftermarket refinish products are available to do this yourself. However, if the main body, bowls, and blocks look crummy due to age and wear, and if you really want them to look correct, a good option is to send the parts to Holley for a proper prep and replating.

Another option is a professional ultrasonic cleaning unit. It offers a great way to clean carburetor parts, using ultrasonic frequencies inside a non-caustic water-based solution. No corrosive chemicals endanger carburetor parts. These units are extremely expensive, so purchasing one is out of the question. However, they may be found at professional shops that specialize in diesel fuel injector rebuilding. ■

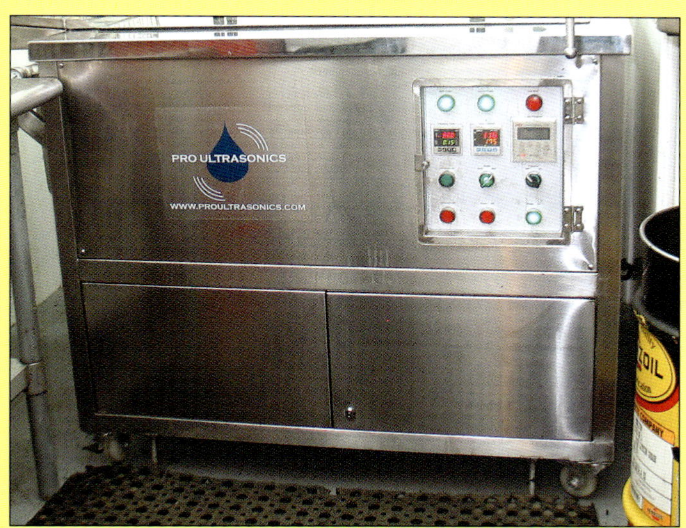

A professional ultrasonic cleaning system effectively cleans carburetor parts, using ultrasonic frequencies inside a non-caustic water-based solution, so no corrosive chemicals endanger carburetor parts. This is one cleaning option that may be locally available to you.

has a tall handle for ease of dunking and removal.

After soaking the parts in solvent, rinse them in hot water to remove chemical residue. Blow dry with compressed air, paying particular attention to blowing out all passages and orifices.

Don't get carried away with air pressure. Adjust the regulator on your compressor to about 25 to 30 psi. Excessive air pressure can potentially dislodge any internal pressed-in plugs, which can ruin your main body and turn it into an expensive paperweight.

Cracks and Pinholes

If you don't have access to expensive flaw-detection equipment, an easy way to check for cracks is with a dye-penetrant kit and an ultraviolet (UV or "black light"). These kits include three hand-held aerosol spray cans: One cleans and prepares the surface, one is a dye-penetrant spray, and one is a developer spray. You can use the following three steps with most kits.

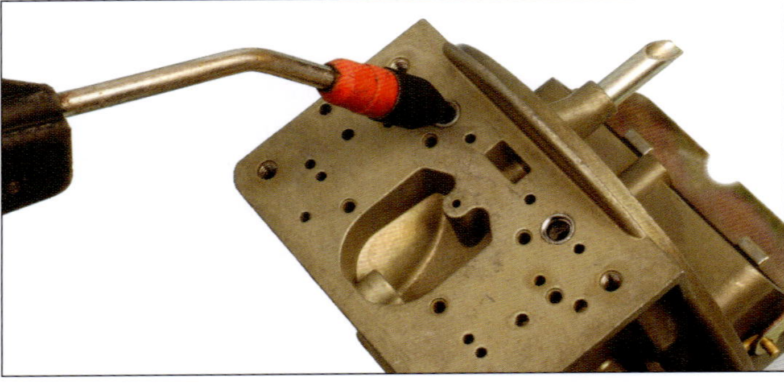

After rinsing the components in carburetor solvent, be sure to blow clean compressed air through all ports, passages, and small orifices to remove any grime, debris, or particles.

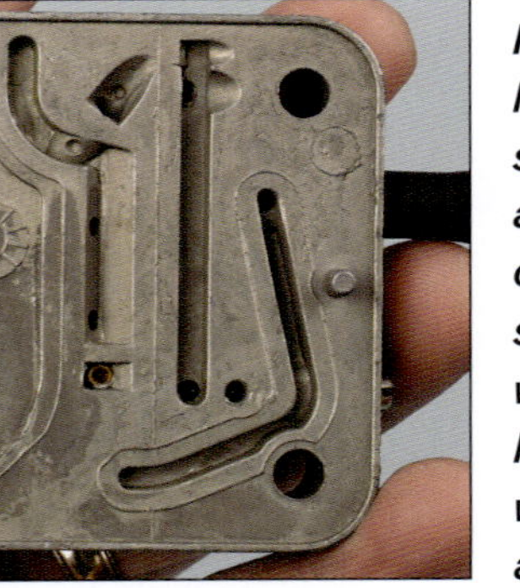

Metering blocks have numerous small orifices. In addition to solvent cleaning, run a small-diameter wire through each hole and follow with compressed air.

A simple aerosol dye-penetrant flaw-detection kit can prove to be handy for inspecting any component that you suspect has a crack or hole. Goodson's kit is an example. First, you spray the surface with the cleaner. This is simply a fast-evaporating solvent that removes any surface contaminants and prepares the surface for the dye. Next, you spray the dye onto the surface and allow it dry for about a minute.

The third step is to spray on the developer. This allows concentrations of the dye to appear, such as in small pinholes or cracks. Because the dye is not visible in daylight or normal room light, a UV light must be used to view the area.

TECH TIP

Media Blasting

Do not sand blast any carburetor part. Also, avoid glass bead blasting. Abrasive blasting can easily result in potential contamination of small orifices, and can possibly ruin precision mating surfaces. The only exception to this might be soda blasting of the main body.

Soda (essentially a refined, purpose-formulated version of baking soda) cleans parts to bare metal; it is not nearly as abrasive as other materials. It dissolves and flushes out easily with a thorough after-rinse. However, soda is only to be used in a dedicated soda-blasting machine. Don't try flushing out a glass-bead machine and dumping in a load of soda. No matter how hard you try to clean the machine, old glass-bead particles may still be present and can contaminate the media. ■

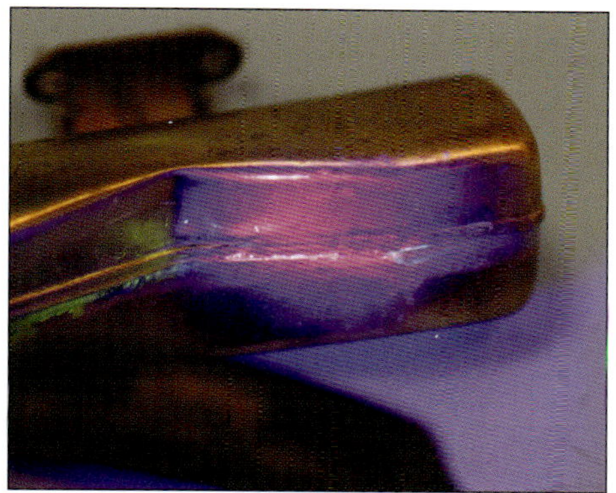

Even with a UV light, inspecting in a brightly lit room may be difficult. It's best to turn off the room lights. The UV light causes any openings, such as cracks or pinholes, to be visible as bright green in color.

A Nitrophyl float (left) must be inspected by weighing it after it is immersed and soaked in solvent or gasoline. A brass float (right) may be weighed or immersed in water to see if bubbles are released. Or, if it was removed recently, you can shake the brass float to try to detect any fuel that may have leaked inside.

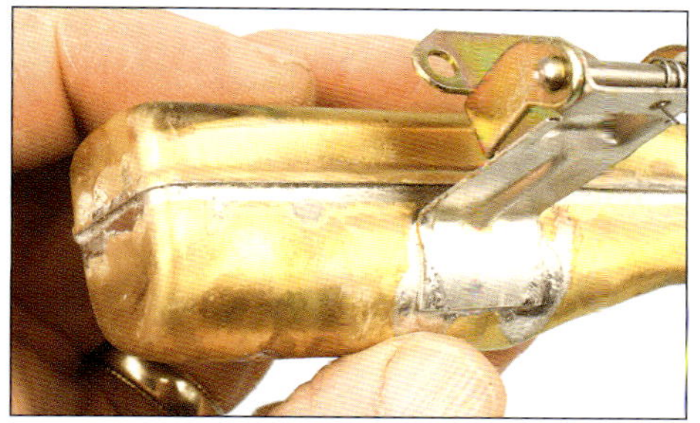

It's rare to encounter a leaking brass float. However, if it leaks, it fills with fuel that prevents it from floating in the bowl. You can immerse the float in clear water to check for air bubbles, or use a dye penetrant to check for body and seam flaws.

Spray the cleaner onto the surface and allow it to dry. Then spray the dye onto the surface so that it penetrates into any existing cracks or pinholes. Finally, spray on the developer.

When viewed under a UV light, any surface flaws or irregularities are revealed in a bright green color.

With regard to carburetors, applications include checking for leaking fuel bowls, metal floats, external bowl-to-bowl fuel transfer tubes, cracked baseplates, and so on. After the inspection is performed, the dye and developer are easily rinsed off with a solvent such as carb cleaner or lacquer thinner.

Another way to perform an easy check to see if a float is leaking (absorbing fuel) is to immerse the float in a clear jar of clear water and look for bubbles. Rotate the float in all directions while it's submerged. If any cracks or holes exist, you should see air bubbles escaping. Any leaking brass floats need to be replaced.

Holley floats may be brass, plastic, or solid Nitrophyl (often called "foam" floats, the material is not foam). If your carb uses a solid float, you must weigh it to determine if it has degraded to the point where it is able to soak up fuel and become heavier. Immerse the float into a container of clean solvent for 15 to 20 minutes, then weigh it.

A center-hung Nitrophyl float should weigh approximately 12 grams; a side-hung Nitrophyl float should weigh about 15 to 16 grams. If you don't have a gram scale, consider visiting a local performance engine builder who balances crankshafts. That shop has a precision gram scale for weighing pistons, rods, bearing, and rings.

Baseplate Inspection

You need to verify the integrity or flatness of the baseplate. If a carb was previously mounted using uneven and/or excessive torque, the baseplate may be warped. Warpage can also occur over time because of a combination of heat cycling (cold to hot or hot to cold) in conjunction with uneven tightening force.

Use a machinist's straightedge and a feeler gauge to check the bottom of the baseplate for flatness: Place the straightedge across the baseplate from corner to corner at the bolt-hole locations, front to rear, side to side, and diagonally. Hold the straightedge in place and attempt to insert a feeler gauge between the baseplate and straightedge. An accepted limit of space is .007 inch. In my opinion, more than .005 inch of warpage should be addressed.

If excessive baseplate warpage is found, you have two choices: Replace the baseplate or have it resurfaced. Do not attempt to do this with any type of power tool and abrasive. You end up with wavy and uneven surface.

If you don't have access to a precision milling machine, you may correct surface warp with a hand file, but the file must be clean and sharp, and the baseplate must be secured in a fixture to prevent it from moving. You may find burrs around the edges as a result of someone prying the carb apart using a screwdriver; you should use a flat file to eliminate these burrs. Be careful not to remove too much material, because it can create issues with clearance at the locating dowel pins on the top side.

If you'd rather not attempt this step yourself, have a local machine shop do it. If the initial measured warpage/distortion is .010 inch or more you should replace the throttle body baseplate.

1 Check Baseplate Flatness

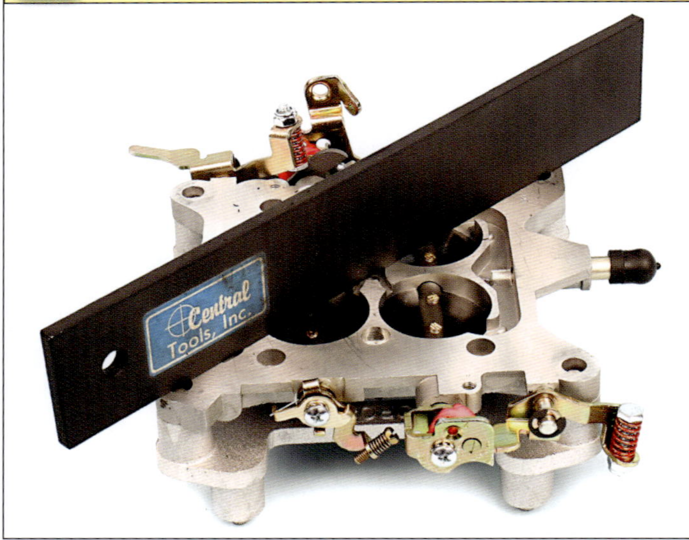

When checking for baseplate flatness, use only a precision machinist's straightedge. Don't rely on a household ruler.

2 Place Feeler Gauge Between Base and Straightedge

Using a feeler gauge, check for gaps between the baseplate bottom surface and the straightedge. Anything beyond .005 inch should be corrected. If the warpage is .010 inch or greater, you should replace the baseplate. Check for warpage from front to rear and side to side.

3 Verify Straightness of Baseplate

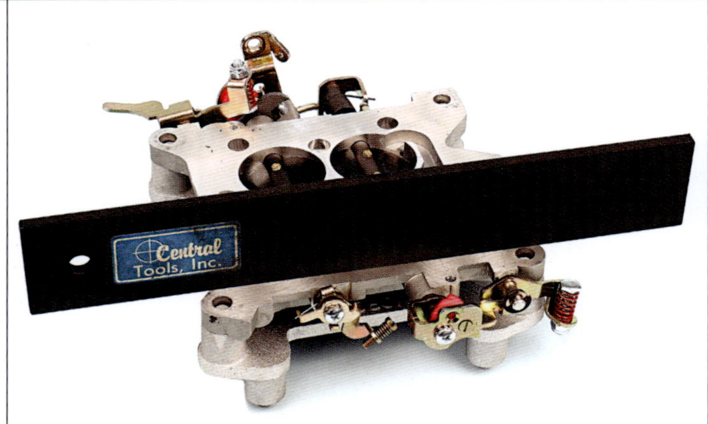

Check for warpage diagonally from corner to corner.

After you confirm that the baseplate is flat (or you install a new baseplate), pay attention when tightening the carb to the intake manifold. Tighten evenly in a crisscross manner and follow the torque specifications.

Main Body

Check the metering block mating surfaces of the main body for flatness with a precision straightedge and feeler gauge. If warpage is found to be more than .007 inch, you can try to resurface the area by carefully using a clean and sharp, broad flat file.

First, make sure that the surface is clean and dry. Then, wipe machinist's dye across the surface and allow it to dry. Using a clean flat file, carefully and lightly file across the surface. The dye is quickly removed on high areas and remains on low areas. Although an ideal surface is perfectly flat (with all dye removed), this really isn't necessary.

As you proceed, check your filing with a straightedge and feeler gauge. A maximum of approximately .006-inch deviation is acceptable because the new gasket should be able to seal with that minimal amount of unevenness.

File very carefully to avoid creating an uneven surface. Don't dwell on one particular area, and make absolutely sure that the file is always held parallel to the surface. Move the file across the entire surface with light strokes.

Also check for flatness at the main body's accelerator pump mounting flange. It's rare for this surface to be warped, but as long as you're checking, it's worthwhile. Again, if using a flat file, the main body must be secured. If the body moves during filing, you may file unevenly and create more problems and the gaskets will not seal properly.

Power Valve

Holley carburetors are equipped with a vacuum-operated power-

Using a precision straightedge and feeler gauges, check the main body metering block decks for flatness. If the surface warped more than .007 inch, it likely requires resurfacing.

Machinist's dye (available in blue or red) is readily available at any shop that sells tool and die and/or commercial fasteners. This Dykem blue is one common type. Machinist's dye is available in both spray cans and bottles. The bottles have a small paintbrush attached to the cap for applying the dye.

Apply the dye evenly and allow a few minutes of drying time before filing. Lay a flat file evenly across the surface. Use both hands and be extremely careful to keep the file flat and parallel to the surface at all times. Tipping the file results in uneven surfacing, potentially ruining the main body.

Initial strokes with a flat file on this main body's metering block deck reveal high spots at each side of the surface. It's not necessary to file until all traces of dye have been removed. As long as the warpage is limited to about .007 inch or less, the metering block gasket should seal properly.

enrichment system that incorporates a power valve. The power valve threads into the primary-side metering block. A power valve in the secondary metering block may also be located on double-pumper carbs with mechanical secondaries. This prevents a lean condition during WOT operating conditions. Placing a power valve at the secondary metering block is of particular benefit if the engine has forced induction, such as a supercharger, where the power valve is able to supply additional fuel to the main system during heavy loads or WOT.

Identification Number

A Holley power valve is always labeled with an identification number, such as 3.5, 35, 4.5, 45, 6.5, 65, which translates into the power valve's vacuum rating. Even-numbered power valves are not available. Some numbers include a decimal point and others don't; the decimal point is irrelevant.

For example, both 65 and 6.5 indicate a vacuum rating of 6.5 inches of vacuum. This is the point at which the valve opens and the engine vacuum drops to that point or below it.

Manifold Vacuum

Holley power valves have a cast metal housing, a rubber diaphragm, and a spring. When engine vacuum drops to and below the valve's rating,

Holley carbs may have one of two styles of single-stage power valves. On the left is a power valve with small orifices. On the right is a power valve with the large window passage. The version with small holes is designed primarily to improve part-throttle economy on vehicles with heavy loads and isn't the best choice for a performance application. The gaskets differ in shape between the two versions. The correct gasket must match the style of power valve.

the valve opens and allows extra fuel to flow through a calibrated port in the metering block, which is called the power-valve channel restrictor.

A blown power valve (from a failed diaphragm) results in an over-rich condition, allowing full-time fuel enrichment. The same problem occurs if the incorrect power valve is installed for the application. This obviously results in lousy fuel economy, spark plug fouling, poor idle quality, and black smoke from the tailpipe.

An unmodified stock engine probably has an engine vacuum at idle in the 18- to 20-inch range. A high-performance or competition engine that's equipped with a long-duration high-overlap camshaft produces a lower engine idle manifold vacuum.

When inspecting for manifold vacuum, be sure to allow the engine to warm up to full operating temperature before taking a vacuum reading. A simple vacuum gauge connected to a full manifold vacuum port may be used when checking engine vacuum.

Before taking a vacuum reading, bear in mind that if the vehicle is equipped with an automatic transmission, the transmission should be placed in gear, with the brake on, and with the engine at idle. If the vehicle has a manual transmission, place the transmission in neutral and check manifold vacuum at engine idle.

Size Selection

How do you determine the size of power valve to use? Well, application has a great deal to do with it.

The instructions from Holley have long been to select a power valve that is half the engine's manifold idle vacuum (in gear for automatics). In other words, if your idle vacuum is 13 inches, your power valve part number should be 65 (rated at 6.5 inches of vacuum).

If, when manifold vacuum is divided in half, the number is an even number, per Holley, always select a power valve with the next lower reading. For example, if manifold vacuum is 12 inches, select a power valve with a rating of 5.5 inches. This has been the baseline recommendation by Holley for a power valve within the general operating range, and it has worked reasonably well, but is it the best approach?

Given that you can have identical idle vacuum readings in a big-block Vega race car or a heavy motorhome, is this procedure specific enough? It's hard to imagine two such dissimilar applications having the same enrichment needs. Although a high-powered, lightweight, hot car rarely sees vacuum dip to 6 inches unless spending a lot of time under full-throttle acceleration, it is highly likely that a heavy, somewhat underpowered application such as a motorhome sees low operating vacuum readings much more often.

Although the Holley method is fine for some applications, using it on street-driven cars can lead to poor fuel economy and part-throttle hesitation. Ideally, you do not want to open the power valve under normal operating conditions or cruise conditions. You want to select a power valve that opens when you need enrichment. The power valve opening point should be as high as possible but below the lowest point within the normal cruising range.

One method for determining the best power valve for a street car is to find the best lean jet for the highest vacuum under cruise conditions, then squeeze on the gas slowly until a lean "stumble" (or surge) occurs, note the stumble vacuum level, and install the next available power valve higher than the stumble vacuum level. This allows for a leaner main jet to be used for best fuel economy and prevents a lean surge at part throttle or light acceleration.

For a real-world example of trial-and-error power valve selection, consider a fairly typical street rod engine that has a manifold vacuum reading at idle of 15 inches. It would be reasonable to suggest under normal operating conditions that the vehicle might see 10 inches of vacuum during normal acceleration and up to 19 inches under deceleration or light-load cruising conditions. With 10 inches of vacuum being the lowest vacuum reading under normal conditions, you want a power valve that opens below 10 inches of vacuum. A 10.5-inch and above power valve opens too soon under normal acceleration, leading to poor fuel economy and a rich condition. The 6.5-inch power valve that many new Holley carbs come with from the factory leave a pretty big hole between the 6.5-inch opening point and the "just under 10" opening point that is closer to ideal.

Using Holley's formula of selecting a power valve rated at half the idle vacuum, the 7.5-inch valve would be the choice, but that still may leave a big enough lean hole to exhibit lean condition issues, including stumble, lean misfire, and ping, just as the 6.5-inch power valve did. That gives you the option of using an 8.5- or 9.5-inch power valve.

Given that mechanical tolerances of manufactured goods, and the varieties of gas available at the pump today, trial and error helps you determine the best valve. If you experience some pinging or detonation,

the 9.5-inch will likely help reduce that issue with earlier enrichment. If you don't have detonation issues, the 8.5-inch valve might give you better fuel economy.

Power Valve Testing

To inspect for a suspected blown power valve, have the engine at operating temperature and idling, then turn the idle-mixture screws (on each side of the metering block) all the way in. If the engine begins to stumble and dies, it is an indication that the power-valve diaphragm is not blown. If the engine does not stumble and die, the power valve is likely faulty.

Power Valve Testing

1 Separate Fuel Bowl from Metering Plate

Use a nut driver or wrench to remove the four screws that fasten the fuel bowl to the carburetor. Once removed, you have access to the metering block.

2 Locate Power Valve and Metering Jets

On the front face of this metering block, the power-valve tip is at the center, flanked by a pair of metering jets.

3 Access Power Valve

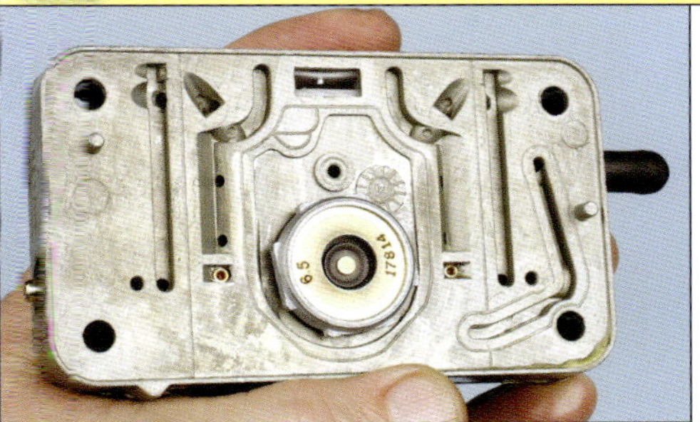

Access the square drive head of the power valve at the rear of the metering block. The metering block is simply sandwiched between the carb main body and the fuel bowl, with no additional mounting screws.

4 Remove Power Valve

The vacuum rating is stamped onto the valve. This example is a 6.5 valve, suitable for an engine that pulls 13 inches of vacuum. A 1-inch wrench may be used to remove or install a Holley power valve with a four-sided profile with opposing flats. Use care with a conventional wrench, because the engagement flats are very shallow at about .120 inch.

When servicing a power valve, use a dedicated power-valve remover/installer tool such as this one from Willy's Carburetors. This type of tool eases power-valve service, eliminating potential slipping and possible damage to either the power valve or the main body. A dedicated power-valve tool is form-fitted to the profile of the Holley power valve.

5 Remove Power Valve CONTINUED

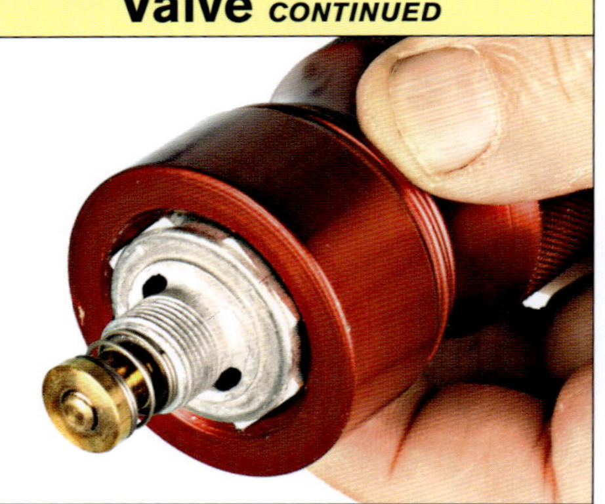

With the specialty power-valve wrench engaged onto the power valve, the recess inside the wrench bottoms-out against the power valve, preventing the tool from contacting the surface of the metering block. The knurled grip area of the tool provides a non-slip surface. This view shows that the wrench securely engages the power valve but does not protrude past the power-valve flats.

6 Inspect Metering Block

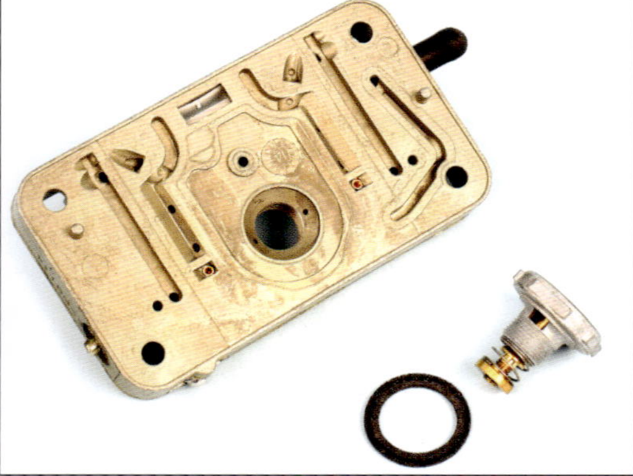

A Holley metering block has a 1/2-27 threaded port into which the power valve mounts. If the female threads in the metering block are damaged and require cleanup, be sure to use a 1/2-27 chaser tap. A chaser tap, in contrast to a cutting tap, re-forms the threads without removing excess thread material. Whenever you remove a power valve, note the shape of the valve gasket and whether the power valve has small holes or large windows. Remember, the incorrect-style gasket results in an internal fuel leak.

7 Inspect Power Valve

The carburetor's power valve threads into the primary-side metering block; it threads into the secondary metering block of Double Pumper carbs with mechanical secondary operation. The power valve supplies additional fuel to the main system during heavy loads or WOT operation.

8 Choose Gasket Carefully

An improperly matched power-valve gasket leaks. Use a gasket with a concentric inside diameter with a power valve that has the large-window fuel opening. Use a gasket with three small tangs on the inside diameter with a power valve that features a series of small fuel holes.

Using a Power Valve Test Tool

Rather than guessing about the condition of a power valve, using this simple tool and a vacuum pump will let you know definitively if the valve is in good condition. ∎

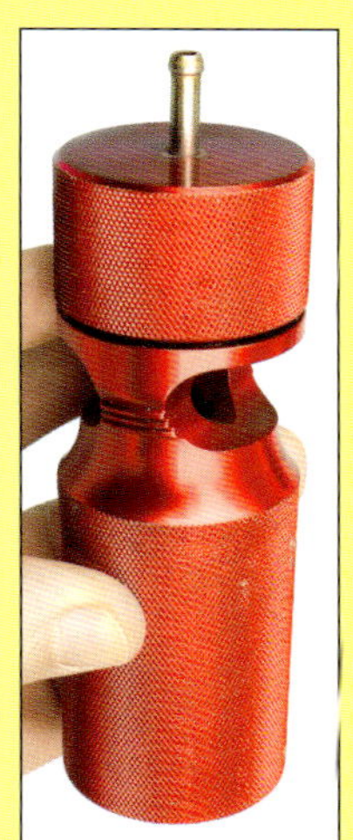

1 *After the power valve has been removed, use a power-valve tester. A specialty tool such as this is available from a number of performance carburetor system suppliers and allows you to easily verify the condition of the power-valve diaphragm.*

2 *This power-valve tester from Willy's Carbs has a head that secures to the tool main body with an O-ring. The main body has a 1/2-27 threaded port that accepts the power valve.*

3 *Wiggle and pull the power-valve tester tool's head off to install the power valve into the tool.*

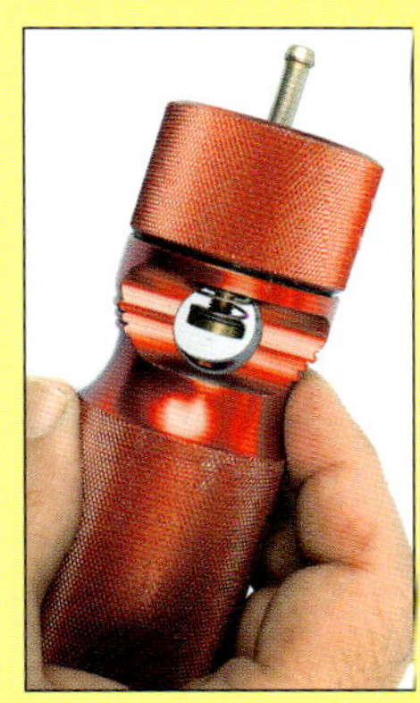

4 *After you thread the power valve into the tool's port, seal the valve to the tool with a separate O-ring. Then, install the tool's head. During testing, an open passage allows you to observe valve movement as vacuum is applied to the tool.*

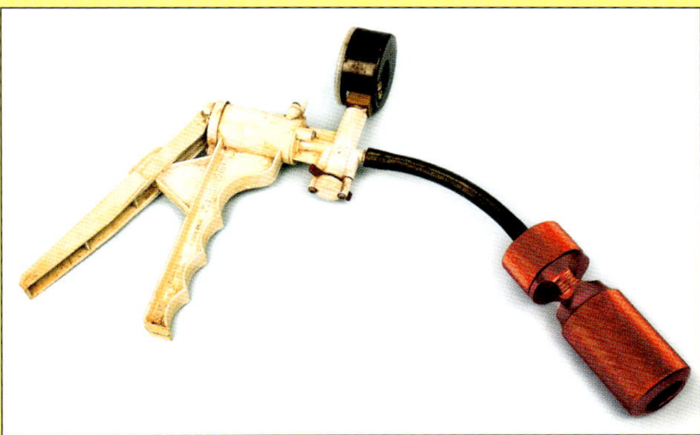

5 *Connect a common gauge-equipped hand-held vacuum pump to the tool's nipple fitting and apply vacuum to test the power valve.*

Power valves are screwed into the metering block, which has a thread size of 1/2-27. The head of the power valve has four flats for wrench engagement and may be removed using a 1-inch wrench. However, the flats are relatively shallow with a height of about .120 inch. Take care to avoid burring the flats or damaging the metering block surface.

An aluminum power-valve wrench is available from select sources; it is designed specifically to securely and safely engage the power-valve flats. If you plan to service Holley street or competition carburetors on a regular basis, I recommend purchasing one of these tools from a source such as Willy's Carburetors.

Another specialty tool is available to vacuum test a power valve with the power valve removed from the carb. These testers are readily available from sources such as Moroso and Willy's Carburetors.

This specialty tool allows you to bench test a power valve with a common handheld vacuum pump. Simply unscrew the head of the power-valve tester tool and then screw the power valve into the tool's main body port. A rubber O-ring seals the tool to the power valve. Reinstall the tool's head, which seals to the tool main body with another O-ring. Connect a hand-held vacuum tester's hose to the 1/8-inch fitting on the tool's head, and operate the vacuum pump.

The pump's gauge allows you to monitor the opening point of the valve. A handy window is built into the aluminum tester tool that allows you to visually check if and at what point the valve opens. You can then determine if you have a faulty power-valve diaphragm and easily verify the level of vacuum when opening occurs. In addition, you can verify the rating of the power valve by observing the vacuum pump's gauge reading as the valve opens. Using this type of testing tool eliminates guesswork and allows you to verify power-valve condition and operation.

Vacuum Secondary Diaphragm

Inspect the power-valve diaphragm closely for damage. An engine backfire can easily rupture this diaphragm, which causes an over-rich condition. In 1992, Holley began to include power-valve blow-out protection that was designed to protect the power valve and prevent diaphragm rupture. This involves a spring-loaded check ball in the throttle body's vacuum port that leads to the power-valve vacuum chamber.

Remove the vacuum diaphragm cover and inspect the rubber dia-

If your car is so equipped, remove and inspect the condition of the rubber. When removed, the diaphragm appears distorted, which is normal.

Inspect the vacuum secondary diaphragm closely, for cracks, splits, tears, or aged rubber and replace if any damage is found.

phragm for tears or pinholes. Replace the diaphragm if you see any damage, including elongation or tears at the screw notches in the rubber. Pay particular attention to the small hole that aligns with the tiny vacuum port on the diaphragm housing; look for tears that travel from the hole in the rubber to the edge of the rubber. If the diaphragm is damaged, hard, or has isolated hard spots (possibly due to age and an extended storage period), replace it with a new diaphragm.

Also inspect the secondary vacuum signal port to make sure that it's not obstructed. Squirting it with carb cleaner and blowing it with compressed air should clear any debris.

Vacuum Secondary Seal

If the carb is equipped with vacuum secondary operation, inspect the vacuum diaphragm housing and cap for flatness. If it has ever been severely overtightened, these mating surfaces may be warped. Measure the flatness of both surfaces with a precision straightedge and feeler gauge. Carefully file it flat if warpage of more than .003 inch is found.

This is a relatively small area, so

You must make sure that the tiny vacuum port in the secondary vacuum housing is clean and free of obstructions. Blow compressed air through the orifice to make sure that air passes through to the vacuum port on the rear of the housing.

The rear vacuum port on the secondary diaphragm housing has a small cork/rubber gasket. If in doubt that it is usable, replace it. A carb rebuild kit includes this. If you have purchased a rebuild kit, it doesn't make sense to reuse the original gasket.

it's best to first apply a light coating of machinist's dye. When the dye has dried, run a fine flat file across the surface gently to reveal any low spots. Carefully file it flat until the surface is uniform.

Unlike metering block gaskets, which are fairly thick and somewhat forgiving, the vacuum diaphragm that seals between the housing and cap is fairly thin. Excess surface warpage can easily result in a leak.

Throttle Plates

Inspect the throttle plates for damage. If the carb was laid on a workbench while someone held the throttle open, for example, the plates may be bent. If the plates are bent, you must replace the choke plates. The staked screws that secure the plates to the shaft or the entire throttle body also need to be replaced. With the throttles closed, there should be a small gap between the primary throttle plates and their throttle bores, as set by the factory. The plates should not be seated fully. They are supposed to be open slightly for idle quality. This is likely in the range of .010 to .012 inch, depending on the model.

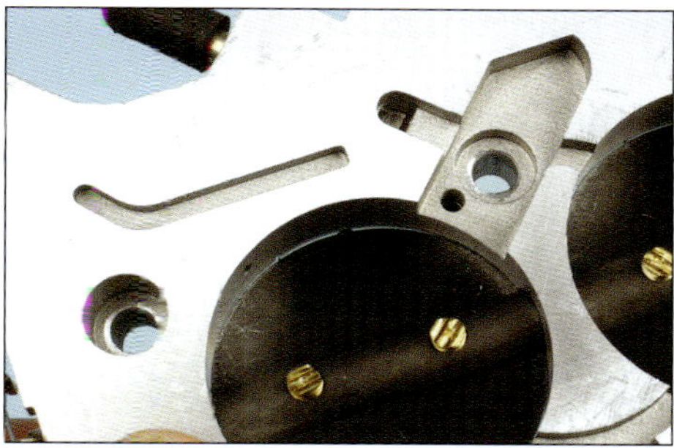

Notice the small gap at the edge of the throttle plate in relation to the throttle bore. This is normal. A gap to keep the primary throttle plates slightly open is set at the factory to aid in engine idle at closed throttle. Depending on carb model, this gap may be in the range of .010 to .015 inch. Do not attempt to modify the gap to fully seat the throttle plates.

Inspect the choke's fast idle cam steps for signs of wear, damage, or burrs. If damaged, replace the choke housing.

Choke Fast-Idle Cam

Inspect the plastic fast idle cam (behind the choke housing) for wear, breaks, cracks, etc. If any of the steps in the cam appear worn or damaged, replace the cam.

Throttle Shaft and Shaft Bearings

Keep in mind that in addition to vacuum leaks, a distorted or warped baseplate throttle body can also result in throttle shaft binding. If the throttle shaft is bent or untrue, it can be especially noticeable in

The temperature-sensitive thermal spring adjusts carb richness according to the ambient temperature so it needs to be in good working order. Inspect the thermal spring inside the electric choke cap. If it is distorted or shows signs of unwinding, replace it. New caps with springs are readily available.

vacuum-operated secondary models because the vacuum diaphragm may not be able to open the secondary throttle plates. If a throttle shaft is bent, replace the shaft or replace both the throttle body and the shaft.

Inspect the throttle shaft (with the carb disconnected from the vehicle's throttle linkage) by wiggling the shaft up and down. Any detectable vertical movement indicates that the throttle shaft bearings are worn.

Wiggle the throttle shafts up and down to check for excess play. If vertical slop is found, the shaft bearings are worn.

This requires that the throttle body be replaced. A competent carburetor rebuilder may be able to re-bush the throttle shaft bore.

Overtightening the carb to the intake manifold, especially when using a very thick and soft gasket, can also contribute to shaft binding.

Idle and Main Well Passages and Emulsion Tubes

Under typical circumstances, preparing to clean a metering block should involve simply removing the idle-mixture screws. When fuel flow through the main system begins, fuel is metered through the main jets and into the main well in the metering block. As fuel travels upward through the main well, air is added from high-speed bleed orifices in the passage to emulsify (pre-atomize) the fuel charge, and then it travels through the discharge nozzles. This air-charged fuel delivery is lighter in weight and is able to respond faster to changes in the venturi vacuum signal.

The tiny air orifices can be clogged if the carburetor was previously subjected to severe contamination. Unfortunately there's no way to verify this without removing the press-fit plugs from the metering block. The top of the metering block has cap plugs that seal the main and idle well passages. Depending on the carb model, removing the press-in cap plugs may reveal the main and idle wells, but some plugs may include an emulsion tube attached to the plug.

To remove the cap plugs and/or plugs with emulsion tubes, carefully drill a 1/8-inch hole in the center of each cap plug. Next, thread in a No. 8 sheet-metal screw into the plug, leaving the head of the screw well exposed above the plug. This gives you something firm to pry against; then, pull the plug out using a pair of small crow's-foot pry bars. Be aware that drilling too deep can easily damage any integral passages that may be present under the plugs and you can easily ruin the metering block. Don't drill more than 1/8-inch deep.

The cap plugs, depending on carb model, may be flat or recessed. If the cap plugs are recessed, carefully measure the installed depth before attempting to remove them because they need to be reinstalled at the same depth. Cupped-type plugs are readily available from Holley (PN 1007-107), but flat plugs are not available. Plugs can be reused if the drilled hole is soldered shut or sealed with an epoxy such as J-B Weld. Care must be taken to avoid excess solder or epoxy from running past the hole and causing an obstruction.

If the carburetor, upon disassembly and inspection, shows signs of heavy contamination, your best recourse may be to simply replace the metering blocks, if you don't feel like taking the time and effort involved in the tedious process of plug removal. If the operation isn't carried out correctly, you can create more problems.

Before installing new or the original plugs, lightly coat the cup outer surface with an epoxy and gently drive the plug into place. Again, if you have cupped plugs, they must be reinstalled at the original depth.

Metering blocks have numerous small orifices. In addition to solvent cleaning, run a small-diameter wire through each hole and follow with compressed air.

Metering Block Disassembly and Inspection

1 **Remove Idle-Mixture Screw from Metering Block**

Prior to cleaning a metering block, remove the idle-mixture screw from each side of the block.

2 **Remove Idle-Mixture Screw Gasket**

Notice the cork-rubber gasket in the idle-mixture screw hole recess. Use a small pick to remove this gasket. Always replace this gasket because strong cleaning solvents damage soft parts such as this.

3 | Measure Metering Block Plug Depth

If you plan to remove a metering block's cap plugs, first measure the plug's installed depth; the new plug must be installed at the same depth. The depth tip of a dial caliper or a dedicated depth gauge (shown) may be used.

4 | Measure Metering Block Plug Depth CONTINUED

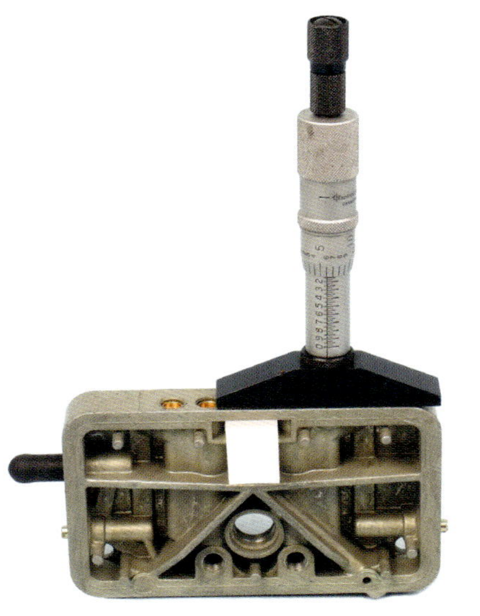

Position the depth micrometer gauge on the metering block top surface. While holding the gauge base firmly onto the metering block, gently turn the fine-adjustment upper black knob clockwise until the gauge anvil stops onto the plug.

5 | Measure Metering Block Plug Depth CONTINUED

A dial caliper has a depth probe at the end of the body opposite the gauge. Extend the probe, and then insert it into the recess of the cap plug.

Carefully insert the dial caliper's depth probe until the caliper body is flush with the surface of the metering block. Remove it and read the depth on the caliper's gauge.

6 | Measure Metering Block Plug Depth CONTINUED

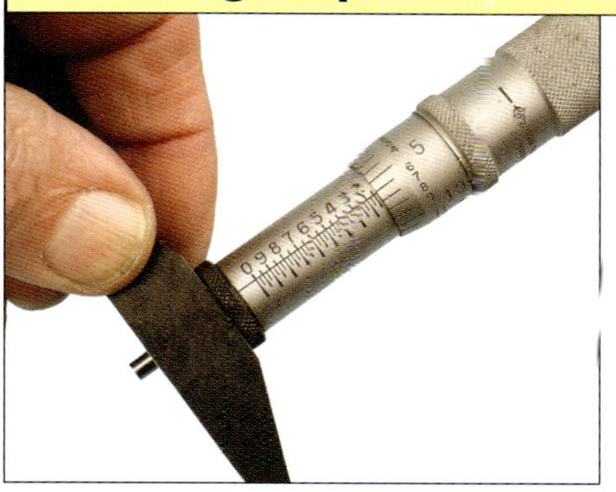

The reading on a sample metering block recessed plug with this depth gauge showed an installed depth of just over .125 inch. If the plug is removed, the new plug must be installed at the same depth.

7 | Drill Out Metering Block Plugs

Main and idle wells are capped with press-fit plugs. To remove them, carefully drill a 1/8-inch hole to no more than 1/8-inch deep. Thread a No. 8 sheet-metal screw into the hole, at no more than 1/8 inch of depth. Use a pair of small crow's-foot pry bars under the screw head to pry the plug out.

Inspect the rubber closely for any signs of damage, such as cracks, pinholes, or aged rubber. If there is any perceptible wear or damage, you need to replace the diaphragm or the pump will leak.

Accelerator Pump Diaphragm

With the accelerator pump cover removed, take out the accelerator pump diaphragm and closely inspect the rubber along the entire perimeter. If any obvious damage such as pinholes, cracks, or tears, is evident, the diaphragm must be replaced. Also check for rubber compliance. If the rubber seems brittle or age-hardened, replace the diaphragm.

Needle and Seat Assembly

Verify that the needle moves freely inside the housing and closely inspect the needle for signs of wear or burrs. Even if you determine that the assembly is okay, always replace the sealing O-ring.

If the tapered needle cone appears worn or damaged in any way, replace the assembly. Remove the rubber O-ring prior to cleaning and make sure that the needle assembly moves freely within its bore.

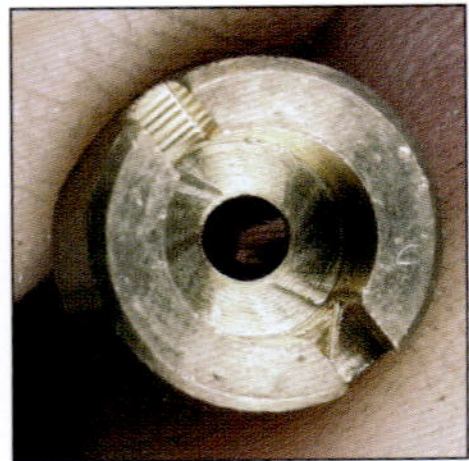

Clean each main jet to verify a clean orifice. A matching-size drill bit can be used, by hand. Never use a power driver to run the drill. Using even the matching diameter drill bit can easily enlarge the hole. If deposits are found, poke and rotate the bit in the passage with only your fingers, followed by solvent cleaning.

Main Jets

As with all other components, the jets must be perfectly clean. Closely inspect the jet orifice to make sure that no debris or varnish buildup is present. The orifice may be cleaned out using a hand drill; never use a power-drill driver. Don't try to enlarge the jet size. If you need larger jets, just install the appropriate size. For cleaning purposes, refer to the list in "Drill Bit Sizes" in Chapter 7 to determine the correct bit size for the corresponding jet. Only use a drill bit for cleaning purposes.

Discharge Nozzles

The nozzle orifices are tiny and must be inspected for cleanliness.

Use a thin wire to clean out the discharge nozzle passages, and blow out with compressed air prior to and after solvent cleaning.

If you have a wire of the appropriate size, run it through the nozzles, clean them with solvent, and then blow them out with air. Under no circumstances should you try to drill the nozzles out. Just make sure that they are clean and free of debris.

Carburetor Renewal Kits

Holley offers an extensive range of renewal kits, available for every model of carb that they make. The kits have everything needed, including the correct small clips, unlike some aftermarket kits that may not be as complete and may not include the correct parts for a specific carburetor. The parts in the Holley renew kits are the same type and quality that was original on the carb from the factory.

Holley offers a rebuild kit (referred to as a "renew" kit) for every model of carburetor that they make. The kits include everything needed to make your Holley carb fresh following a disassembly. Avoid non-Holley brand kits, as they might not include everything, and may not have the same quality of components. This is a renewal kit (PN 37-485) applicable to 4150 carbs in 600, 650, 700, 750, 800, and 850 cfm.

ASSEMBLY

When you are ready to reassemble the carburetor, all parts must be clean and free of any contaminant and gasket residue. Rather than guessing at fastener tightening, use a calibrated in-lb torque wrench and follow all torque specifications.

Your workspace should be large and clean. Carefully organize all parts prior to assembly. This is time well spent and makes the assembly process proceed more smoothly if you don't have to waste time searching for parts. Organize the parts in a logical order, starting with the throttle body and main body at the center, with all primary-side parts organized at one end and secondary-side parts at the opposite end. Carefully place all new gaskets and clips next to the associated parts, and so on.

To prevent small parts from accidentally rolling off your workbench, consider placing a clean towel on the bench first. Because a towel isn't smooth like a wood or metal workbench, the small parts are less likely to roll or slide away.

Lubricant Application

Be sure to lightly coat all rubber O-rings with a rubber-safe lubricant, such as Vaseline. It's a good idea to lightly lubricate all screw threads, to ease installation, and to reduce the chance of thread galling. ■

Main Body to Throttle Body Baseplate

Position the main body upside down on a clean workbench; be careful not to damage the vent tubes or choke plate housing. It's a good idea to place a clean, soft rag or towel on the workbench to avoid damaging the top surface of the main body. Install a new throttle body gasket to the throttle body; align the two small locating dowel pins on the top of

On a clean workspace, begin assembly with the clean main body and throttle body.

the throttle body with the matching holes in the gasket.

Mate the throttle body to the main body carefully; align the throttle body's dowel pins to register to the bottom of the main body.

Install and tighten all six 12-24 x 3/4–inch screws and torque 30 to 50 in-lbs in a crisscross pattern to evenly distribute the clamping force.

Main Body Assembly

1 Install Main Body Gasket

Position a new throttle body-to-main body gasket. The gasket registers on two small dowel pins. Do not apply any adhesive; install the gasket dry.

Torque Fasteners

2 Torque Main Body Screws

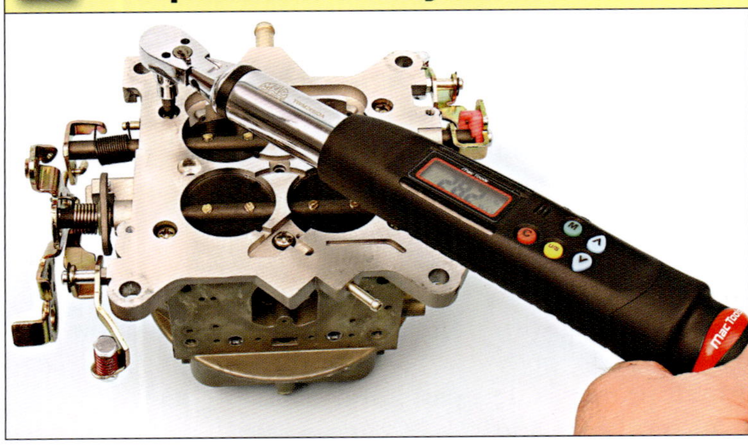

Tighten fasteners with a 1/4-inch-drive in-lb torque wrench. Any style of calibrated torque wrench is acceptable. This is a digital model from Mac Tools. You simply scroll to the desired torque value, which is displayed on the screen, and press a button.

To avoid warping the throttle body, tighten each throttle body-to-main body screw in a crisscross pattern to distribute the clamping load. Final torque is 30 to 50 in-lbs. This digital torque wrench has green LED lights that illuminate progressively as you approach the desired torque; at the selected torque, the lights change to red and the wrench provides an audible beep.

3 Torque Main Body Screws CONTINUED

All of the throttle body-to-main body screws should be below-flush with the bottom of the throttle body base.

4 Mount Carburetor to Work Stand

With the main body secured to the throttle body, mount the assembly to a carburetor stand to make assembly easier.

Metering Block Assembly

Using parts from a model-appropriate Holley renew kit, install new cork rubber gaskets into the metering block's idle-mixture-screw hole cavities. To avoid tearing the gaskets, do not use a sharp instrument to push the gaskets into place. Carefully thread the idle-mixture screws into place and gently tighten them until they bottom out, then back off the screws by about 1½ turns.

Install the power valve and a new gasket into the metering block. Two power-valve designs and related power-valve gaskets are available. It's important to use the correct gasket; using the incorrect gasket results in a fuel leakage around the power valve.

Power valves that have multiple drilled fuel holes must use gasket PN 8R-669. The outer diameter of this gasket is round and it also has a series of three small tangs on the inside.

Power valves with two rectangular window openings must use gasket PN 8R-1597. Both the inside and outside of this gasket are round and it has no inside tangs.

Use a 1-inch box wrench or a dedicated power-valve wrench to install the power valve to the metering block. A specialty billet-aluminum power-valve wrench provides enough engagement of the four-flat-side power-valve head to drive the head without coming in contact with the surface of the metering block. This eliminates accidental scuffing or gouging of the metering block. Using a 1-inch box wrench is certainly acceptable, but you must be careful not to dig the wrench into the metering block.

1 Install Power Valve

Thread the power valve into the metering block from the rear of the metering block. Make sure that you install the correct style of power-valve gasket based on the design of your power valve.

3 Install Main Jets

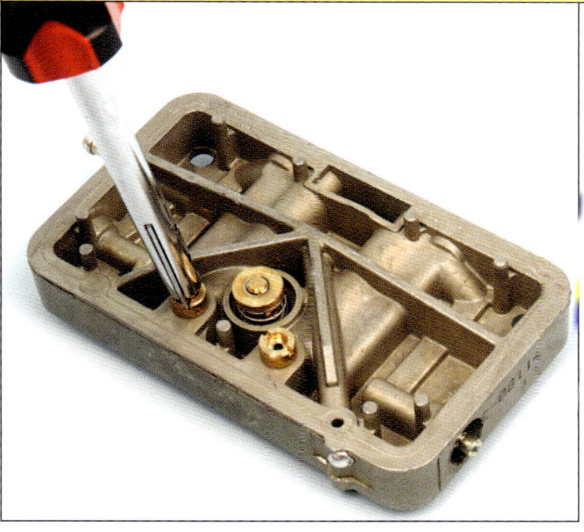

Install the main jets in the metering block; use a dedicated jet driver instead of a conventional screwdriver. A Holley jet driver safely captures the jet and eliminates accidental slippage or burring of the jet's drive slot. The tightening specification for main jets is 30 to 40 in-lbs. Because you can't use a torque wrench along with the jet driver; bear this torque value in mind. Tighten it snugly but don't get carried away.

Special Tool

2 Install Power Valve CONTINUED

With the power valve finger-tight, snug the power valve with a 1-inch wrench or a specialty power-valve wrench (shown). This tool properly engages the shallow flats of the power valve, but it does not engage too deeply and therefore prevents marring the metering block surface. After you install the power valve, use a torque wrench to tighten to 40 to 50 in-lbs. Again, be careful not to dig the socket wrench into the metering block.

4 Install Vent Whistle

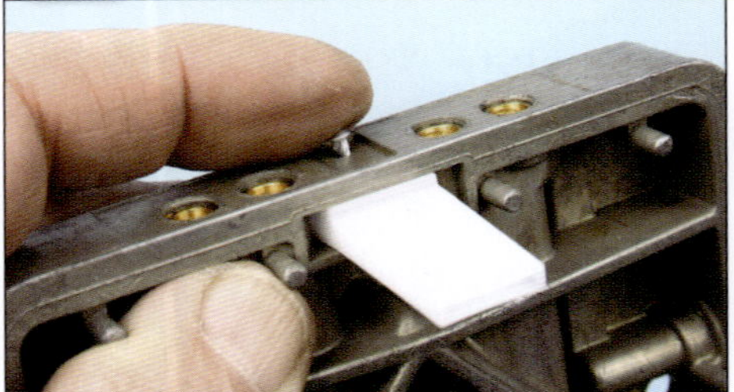

If your metering block was equipped with a vent whistle, install it now. It slips into the top cavity in the metering block with the whistle vent facing the fuel bowl side. Insert a new tap-in rivet pin through the top of the metering block.

Tap the rivet pin with a plastic hammer or the grip butt of a screwdriver until the pinhead is flush against the metering block.

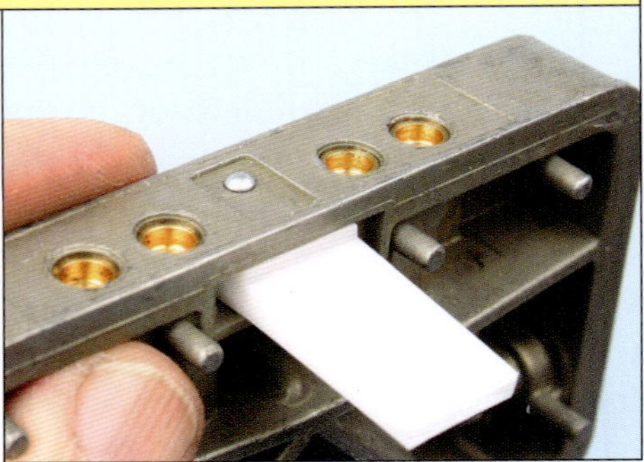

The pin simply locates the vent whistle and prevents it from dislodging. Set the assembled metering block aside until you're ready to install the fuel bowl.

Accelerator Pump Installation

In this procedure, you are going to install the accelerator pump to the fuel bowl. Only a few parts are required, but these are sensitive parts so this requires a light touch and attention to detail.

1 Organize Accelerator Pump Components

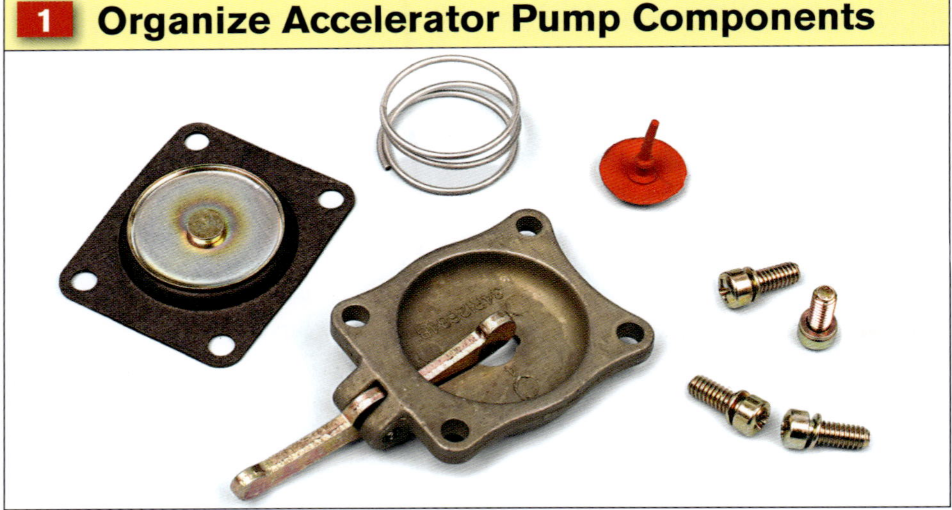

These parts (clockwise from left) are the accelerator pump diaphragm, accelerator pump diaphragm spring, rubber check valve, housing screws, and accelerator pump housing.

Important!

2 Install Check Valve

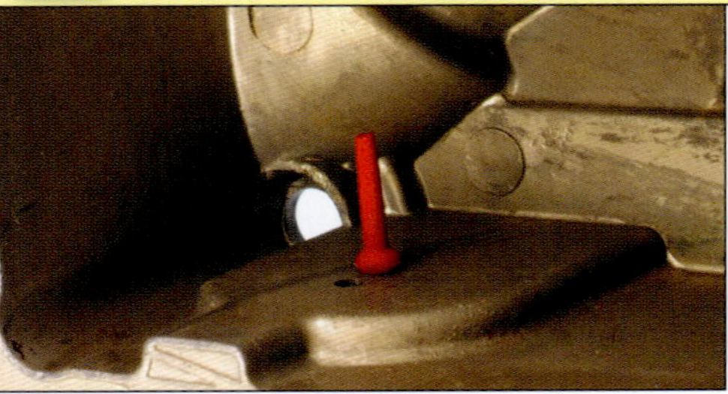

! *Install a new rubber check valve from a renew kit. Handle it with care to avoid tearing.*

Lightly lubricate the stem of the check valve and pull the stem through the hole in the bottom of the fuel bowl (at the center of the accelerator pump cavity). Pull gently until the stem's barb has passed into the interior of the bowl.

3 Install Check Valve CONTINUED

Using snips or a razor blade, trim off the excess of the check-valve stem to provide clearance for the float.

4 Install Check Valve CONTINUED

Here, the stem has been trimmed; avoid cutting it excessively. The barb must remain to retain the check valve in place.

Here's a new check valve installed, viewed from the bottom of the fuel bowl. Make sure that it is fully seated and not distorted.

5 Install Diaphragm Spring

Important!

With the diaphragm inverted, install the accelerator pump diaphragm spring into the recess (not shown). When installed, the diaphragm must be oriented with the center button facing away from the fuel bowl. This is the contact point for the pump lever. The side of the diaphragm with the shallow rivet head must face toward the recess in the fuel bowl.

6 Install Accelerator Pump Diaphragm

Place the acceleration pump diaphram in position on the housing, contacting the spring. Carefully align the screw holes of the diaphragm gasket to the threaded holes in the fuel bowl.

7 Install Accelerator Pump Cover

Place the accelerator pump cover and lever assembly in place, start all four 8-32 screws by hand, and then torque all four screws to 5 in-lbs. Be sure that the linkage arm captures the accelerator pump arm.

8 Install Accelerator Pump Cam

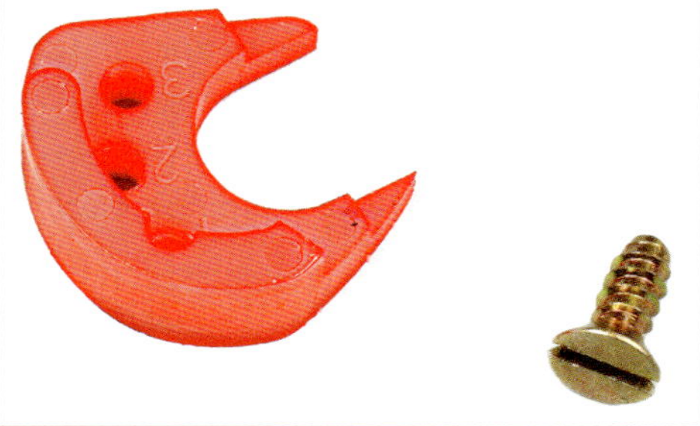

An accelerator pump cam, installed on the linkage, controls the opening rate of the accelerator pump. The mounting screw has three positions; three opening rates are possible. The specific carb model's specifications dictate the factory-recommended position. A single screw secures the cam.

9 | Set Clearance of Accelerator Pump Arm and Lever

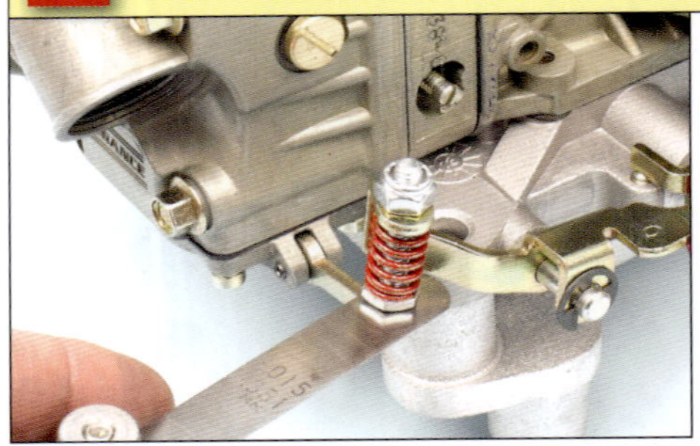

Initially, adjust the linkage lever at closed throttle to obtain zero clearance between the lever and the accelerator pump arm. Then, move the throttle to wide open. At WOT, check that you have approximately .015 inch of additional travel for the accelerator pump diaphragm arm. To adjust the travel, hold the bottom hex nut while turning the top hex to shorten or lengthen the adjuster's spring. If you adjust the throttle to a lower idle speed, it moves the pump cam farther away from the lever, which delays the accelerator pump shot. Deal with this adjustment once the carb is mounted to the engine.

Float Installation

Depending on carburetor style, adjustment of the float may be made internally or externally. For a float bowl design that provides no external fuel level adjustment, the float must be adjusted by measuring the distance from the float to a reference surface, usually the bottom of the float bowl. This must be done with the float bowl removed from the carb. For designs with external adjustment, the float level can be adjusted without disassembly.

After removing the fuel-bowl sight plug, loosen the locking screw on top of the bowl; this allows the float level to be adjusted. Simply turning the hex nut causes the needle and seat assembly to raise or lower. Some models have a clear sight glass, so removal of the sight plug is unnecessary. Adjustments must be made with the engine running.

Center-Hung Float

The two screw holes on the float bracket align with two threaded holes in the upper area inside the bowl. The pivot pin of the float bracket slides into two side grooves that are below the screw holes. Align the screw holes and use a small flat-blade screwdriver to install the two mounting screws.

With a center-hung brass float, an initial adjustment can be made during carb assembly. With fuel in the bowls, adjust the depth of the needle and seat assembly on the engine. With the float, needle, and seat installed, invert the bowl. To adjust the needle and seat assembly, turn the needle's adjuster nut until the top surface of the float is parallel to the roof of the fuel bowl. Remember that you're looking at it upside down.

With the float adjusted initially, you should just be able to see the two screws that secure the float's bracket to the bowl. This provides a gap of about 5/16 inch between the top of the float to the roof of the bowl. This will get you into the ballpark; final float adjustment is made on a running engine.

The Holley Renew Kit's instructions provide recommended clearance dimensions, based on the specific carb model. A primary dry-float adjustment may be 13/16, 3/8, or 11/32 inch. Secondary dry-float adjustment may be 3/4 to 1/2 inch, again depending on the carb model. The renew kit comes with a handy ruler so you can accurately measure float depth.

While holding the nut secure, use a large flat-blade screwdriver to loosen the locking screw. The object is to adjust the float level to a point at which the fuel level in the bowl just begins to dribble out of the bottom of the sight hole. Rock the engine or vehicle from side to side to slosh fuel back and forth in the bowl. This provides a starting point.

Start the engine and adjust the needle until the fuel level is at the bottom of the sight hole, just barely dribbling out. Stop the engine. Hold the adjuster nut stationary and tighten the locking screw.

Fuel Float Installation

1 | Inspect Float

When installing a center-hung brass float, make sure that the pivot pin is centered; an equal amount of the pin should be exposed on each side.

2 Install Fuel Float Bracket

Attach the center-hung float bracket via the two threaded holes inside the bowl. The float's pivot pin seats inside the notches, which are located at the mid-point of the float bracket recess.

Insert the float bracket into the recess with the pivot pin seated in the notches. Secure the bracket with the two 6-32 x 1/2–inch screws and tighten to 5 to 7 in-lbs.

3 Verify Movement of Fuel Float

Check for free vertical movement of the center-hung float; it should move easily. With the fuel bowl inverted, the top surface of the float should be parallel to the roof of the fuel bowl and you should just be able to see the float's bracket screws.

Initially, adjust the depth of the needle and seat assembly to achieve a gap of about 5/16 inch between the float and bowl roof.

4 Install Needle and Seat Assembly

Prior to installing the needle and seat assembly, lightly lubricate the threads and the O-ring with a rubber-safe lube such as Vaseline. This helps to protect the O-ring from damage. Carefully insert the needle and seat assembly into the threaded port on top of the fuel bowl. The upper threads on the brass valve body should engage easily by hand.

5 Thread Adjuster Nut onto Valve

Install a new gasket between the fuel bowl and adjuster nut, and then thread the adjuster nut onto the valve by hand.

6 Install Gasket on Locking Screw

Install a new gasket on the locking screw and thread the locking screw into the brass valve by hand.

7 Adjust Fuel Float

Use a 5/8-inch box wrench on the adjusting nut of an externally adjustable float to hold the nut stationary and prevent it from turning. While holding the adjuster nut in place with the wrench, loosen the locking screw, and then turn the nut to raise or lower the float level. Turning the adjuster nut clockwise lowers the float level and turning the nut counterclockwise raises the float level.

Before adjusting, remove the sight plug from the side of the fuel bowl. Place a rag under the fuel bowl to catch any fuel that may spill out of the sight hole. On a running engine, make the final float adjustment by removing the fuel bowl sight plugs.

Side-Hung Float

When installing a side-hung float that offers no external adjustment, carefully bend the mounting tab for final float adjustment. Because of restricted access after installation, float adjustment must be made prior to installing the fuel bowl. With the fuel bowl inverted, adjust until the top of the float is parallel to the roof of the fuel bowl. Refer to the Holley Renew Kit for the specific dimension for your carb part number.

With the bowl inverted, side-hung floats can also be adjusted to a specific distance between the top of the float and the roof of the bowl. The primary float can be adjusted to achieve a 7/64-inch gap between the toe of the float (the portion of the float farthest from the pivot point) and the bowl roof. You can use a 7/64-inch drill bit as the gauge.

Metering Block Installation

The secondary float may be adjusted for a 13/64-inch gap between the heel of the float (the area of the float closest to the pivot point) and the roof of the bowl. You can use a 13/64-inch drill bit as a gauge.

Install the front metering block-to-fuel bowl gasket onto the metering block. Again, register the gasket onto the locating pins on the metering block. The secondary-side metering plate on a 4160 installs independently to the main body with 8-32 x 1/2–inch clutch-head screws torqued to 12 to 18 in-lbs. It fastens to the main body separately; it is not held in place by the secondary fuel bowl.

Critical Inspection

1 Inspect Locating Pins

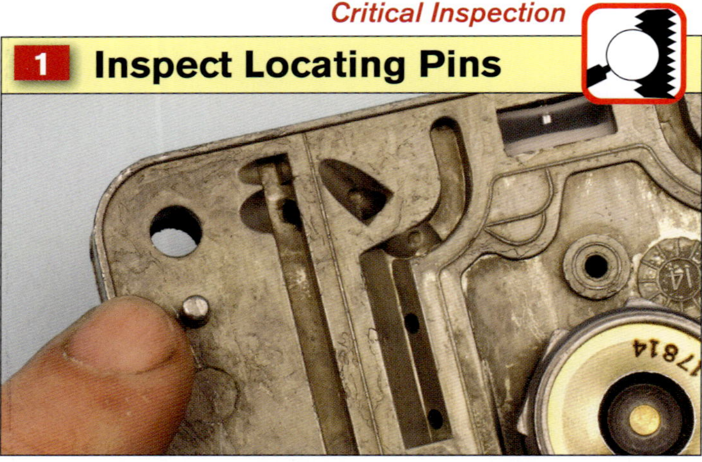

The locating pins on the rear of the metering block are close to the sides of the block. The two small locating pins on the metering block register into two small pin holes in the face of the main body.

2 Inspect Locating Pins CONTINUED

The locating pins on the main body are in the middle of the casting. When assembling the carb, place the metering block locating pins in these holes found in the main casting.

3 Install Air-Mixture Needle in Metering Block

Be sure to install a new cork-rubber gasket in the metering block's mixture-screw port. Lightly lubricate the mixture screw and begin thread engagement by hand. Be careful to avoid cross threading. Continue to gently tighten the mixture screw with a small flat-blade screwdriver until it just bottoms-out, and then back it out about 1½ turns.

4 Install Gasket on Metering Block

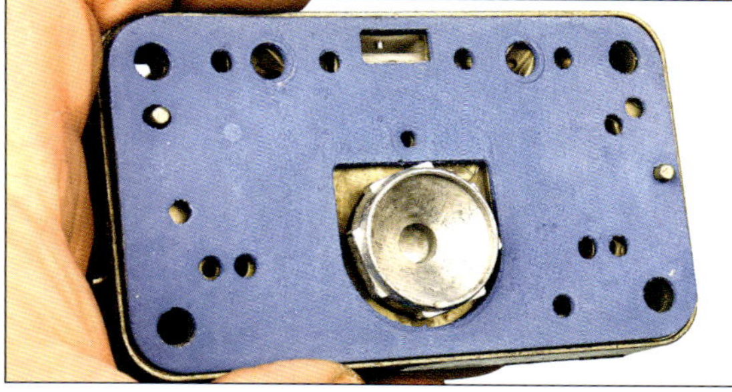

Place a new metering block-to-main body gasket onto the back side of the metering block and register the gasket to the two small locating pins on the metering block.

5 Install Gasket on Metering Block CONTINUED

Install a new front gasket onto the face of the metering block, registering the nine holes in the gasket to the nine locating pins on the front of the block. Place the metering block onto the main body; make sure that the rear gasket remains registered to the block. Do not apply any adhesive to either gasket.

To make registering it with the main body a bit easier, and avoid having the gaskets slip out of position, hold the block with the mating deck upright (vertical). Hold the metering block in place to prevent it from accidentally dropping off the main body. The metering block does not screw onto the main body. It is sandwiched in place, between the main body and fuel bowl, with the four fuel-bowl screws.

Fuel Bowl Installation

Install a new nylon gasket onto each fuel-bowl screw. With the metering block positioned on the main body, place the fuel bowl onto the metering block while lifting the accelerator pump throttle linkage (to allow the accelerator pump arm to contact the underside of the throttle arm). Insert the four 12-24 fuel-bowl screws and finger-tighten to engage the threads.

When all four screws are engaged to the female threads in the main body, keep the fuel bowl against the metering block and tighten the four bowl screws; torque to 25 to 30 in-lbs, using a crisscross pattern. When adjusting the accelerator pump arm contact, fine-tune the length of the spring-loaded adjuster bolt until you have zero clearance between the adjustable spring contact linkage and the pump lever at closed throttle.

Then, move the throttle linkage to WOT. At WOT, use a feeler gauge to make sure that you have about .015-inch clearance for additional travel of the diaphragm lever.

Carbs with an External Fuel Transfer Tube

Some carbs, such as Series 4160 (PN 1850), have an external fuel transfer tube that connects the front fuel bowl to the rear fuel bowl. A rubber seal at each end of the transfer tube seals the tube to the fuel bowl.

Fuel Bowl Screw Length

Holley fuel-bowl screws have a thread diameter of 12 and a thread pitch of 24 (12-24 machine screws). However, be aware that screw shank length differs depending on the distance between the carburetor main body and the outer face of the fuel bowl. For example, because a 4160-series carb has a thin metering plate at the secondary side instead of a metering block, the secondary bowl screws are shorter than those used for the primary bowl. ■

Carb	Screw Length (inches)
4160 primary bowl	2½
4160 secondary with metering plate	1⅞
4150 primary and secondary bowls	2½
4500 primary and secondary bowls	2½

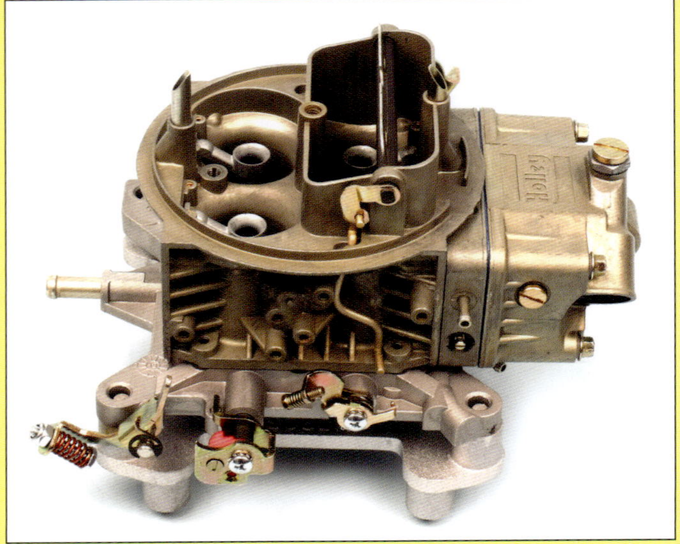

Here, the primary metering block and fuel bowl are installed on this Double Pumper.

Before installing the fuel-bowl screws, install a new nylon washer on each bowl screw. With the metering block held in place, install the fuel bowl and finger-tighten all four fuel-bowl screws. Before installing the fuel bowl, be sure that the accelerator pump linkage on the throttle body is held upward, so that the accelerator pump arm contacts the bottom of the linkage arm's spring-loaded pump adjuster.

Install new rubber seals to the primary and secondary fuel-bowl ports. Lightly lube each end of the transfer tube with Vaseline. Insert the fuel transfer tube into an already-installed fuel bowl.

Install the opposite fuel bowl; engage the free end of the fuel transfer tube. Slide the fuel bowl into position while allowing the fuel tube to slide into its rubber seal. When it is fully installed, the bulge at each end of the tube should be seated against its rubber seal.

Discharge Nozzle Installation

When installing a discharge nozzle, be aware that the small accelerator check-valve pin must be dropped into the nozzle well first and that it uses two small gaskets. Using a hemostat when handling small components, such as placing the pin and the gaskets, is extremely helpful. Place a drop of lubricant, such as Vaseline, onto the lower gasket to help hold it in place. When placing the discharge nozzle, make sure that the notched area engages with the guide tang before attempting to tighten the nozzle's mounting screw.

1 Install Accelerator Check-Valve Pin

Before installing the discharge nozzle, remember to reinstall the accelerator check-valve pin. Drop the pin into the discharge passage with the pointed end down. This pin is being placed into the secondary side.

To ease handling and installation of the very small accelerator check-valve pin in the primary side of a carb that's equipped with a choke plate, it's helpful to secure the pin with a hemostat because the access is fairly tight. Locate the tip of the pin into the discharge passage and then carefully release the hemostat's grip, allowing the pin to drop into place.

2 Install Discharge Nozzle

Before installing the discharge nozzle, make sure that the nozzle and its mounting screw are clean. Always use a new lower gasket and new upper metal washer. When the screw is tightened, the metal washer deforms into a concave shape to seal the countersunk mating between the screw head and the top of the discharge nozzle.

Apply a light coating of lube, such as Vaseline, to the discharge nozzle's lower gasket and place the gasket onto the bottom of the nozzle. This eases installation and helps to prevent the gasket from slipping. Position the upper metal washer over the screw and then install the screw; tighten it to 30 to 50 in-lbs. Make sure that the center notch on the discharge nozzle body seats into the small pocket on the main body.

Choke Installation

A choke mechanism incorporates a plate that closes over the primary side to create a stronger vacuum signal to restrict air and allow a richer fuel mixture to help a cold engine start. The choke plate remains closed or nearly closed to increase fuel atomization until the engine warms up. This vacuum-assisted choke also has a rubber diaphragm that opens the choke a little as engine vacuum begins to develop when the engine starts; it helps to prevent an over-rich condition. The choke plate works in tandem with a fast-idle cam that increases engine speed during this warm-up period.

Most Holley carbs with a choke use either a mechanical or an electric choke. The mechanical choke is controlled directly by the driver via a cable. The electric choke is fitted with a thermostatic spring inside a housing; careful adjustment provides dependable cold-engine starts.

Manual Choke Installation

1 Install Choke Housing

Before installing the choke housing, rotate the spring-loaded choke arm so that the small arm rests against the actuator pin (shown). As you install the housing into the main body, the hole in the long arm must engage onto the choke-plate rod.

Secure the choke-housing screws to the main body, making sure that the choke-plate rod is engaged in the lever arm. Torque the screws to 6 in-lbs.

2 Install Cotter Pin on Choke Rod

The choke-plate rod protrudes past the spring-loaded choke-housing lever. A cotter clip must be installed to the hole in the end of the rod.

Using a hemostat to install the hairpin clip is advisable because there probably isn't enough access for your fingers.

Install the hairpin clip onto the choke-plate rod.

Electric Choke Installation

1 Install Electric Choke Assembly

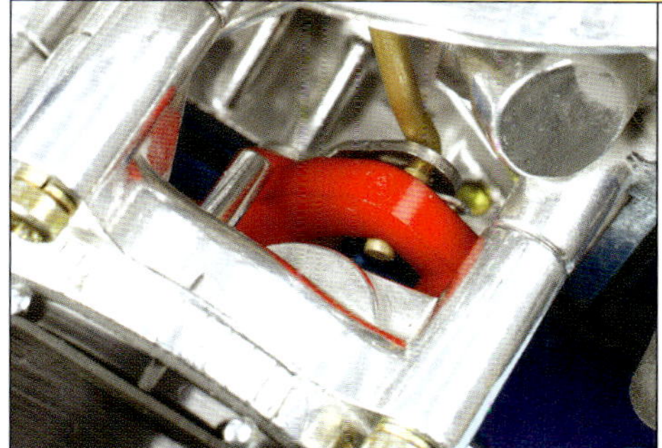

Slide the electric choke until the choke plate rod is captured by the fast-idle cam. You then need to secure the rod to the arm.

A vacuum port is at the rear of the electric choke housing. Be sure to install a new cork-rubber gasket.

Install a new gasket between the electric choke housing and cap.

2 Install Electric Choke Assembly CONTINUED

Make sure that the eyelet on the wound spring engages the lever tang in the choke housing when you install the electric choke cap.

Install the spring washer onto the choke cap and lightly secure the three mounting screws in the cap.

3 Adjust Choke Assembly

Initially, adjust the choke cap index at the center mark on the choke housing. The choke setting can be fine-tuned on the engine.

Vacuum Secondary Installation

The vacuum secondary system contains a vacuum housing with a rubber diaphragm and spring. It's important to use care during assembly to prevent distorting the diaphragm, which will cause an immediate vacuum leak and inoperative secondary actuation. Assembly is a bit easier if the housing is held upside-down while installing the spring, diaphragm, and housing cover.

1. Install Vacuum Secondary Diaphragm

It may be difficult to avoid a vacuum leak while installing the secondary diaphragm into its housing. Try to avoid installing the diaphragm while the rubber is "tulip" shaped; it is difficult to seat the rubber properly.

Push the rubber into its proper shape, with the gasket surface flat where it contacts the housing. This helps to flatten out the rubber.

Notice the small vacuum orifice at the upper right. The captured hole in the gasket must align with this orifice.

2. Install Accelerator Pump Gasket

Push the diaphragm into the cap while you hold the cap, spring, and diaphragm upside-down on a flat surface (this keeps the rubber flat). While holding the cap and diaphragm, slip the housing over the rod and onto the rubber, aligning the screw notches of the gasket to the screw holes in the cap and housing. It's a bit tricky, but be patient; you'll get it.

3. Install Gasket into Vacuum Port

With the cap secured by the four screws, examine its perimeter to see if the gasket has pushed out. If all looks good, you can perform a vacuum check. Be sure to install a new cork-rubber gasket in the vacuum port.

Important!

4. Seat Rod in Secondary Vacuum Assembly

! While holding a finger over the vacuum port, push the rod into the housing until it is fully seated; release the rod while still plugging the vacuum port. If the rod drops freely, you have a vacuum leak and must re-position the diaphragm.

5. Install Secondary Vacuum Assembly

Install the secondary vacuum assembly onto the main body with its three screws.

6 Install Secondary Vacuum Rod onto Pivot Lever

Slip the eye loop of the vacuum rod over the pivot lever.

7 Install C-Clip over Pivot Lever

Install the tiny C-clip to secure the bottom of the operating rod; using a hemostat can make this a bit easier. The clip is very small, so be sure to handle it over a clean work surface in case you drop it. Verify that the C-clip is fully seated in its groove.

TECH TIP — Why Use a Torque Wrench?

During assembly, use the torque specifications listed below for the various fasteners. Although assembly may be accomplished by "feel," it leaves too much room for error. Insufficient tightening can lead to fasteners vibrating loose, as well as potential vacuum and/or fuel leaks. Overtightening can result in stripped threads, component warpage, and possible cracking of cast components. Unequal tightening can lead to warpage and leaks.

As long as you're not in a rush (rebuilding a carb trackside to start a race, for example), take the time to do it right. Tighten all fasteners, where practical, with a calibrated 1/4-inch-drive in-lbs torque wrench. Never be in a rush when you are using a torque wrench. Avoid yanking on the wrench. Tighten by pulling smoothly as you approach the desired torque value and stop as soon as you reach the value. Don't tighten any further. ∎

Fastener Torque Specifications

Fastener	Spec (in-lbs)
Fuel-bowl screws	25–30
Secondary metering-plate clutch-head screws	12–18
Adjustable needle/seat lock screw	50–60
Center-hung float screws (6-32 x 1/2)	5–7
Fuel inlet valveseat (5/16-24)	60–80
Fuel inlet valveseat (7/16-24)	70–90
Power valve	40–50
Accelerator pump cover screws	5
Discharge-nozzle screw	30–50
Baseplate to main body	30–50
Choke-plate screws	6
Fast idle cam nut	10
Secondary shaft nut	10
Throttle-plate screws	10
Main jets	30–40
Carb base to intake manifold	60–80

Throttle Plate Installation

The following is an overview of the procedure for installing a throttle plate.

Step 1: With the throttle-plate screw holes aligned with the shaft, install the screws. Add one drop of Loctite 290 to each screw thread. Snug down both screws, and then back them out half a turn. Lightly open and close the throttle. Do this at least six times so the plate can seat in the throttle bore.

Step 2: Tighten the screws. Open and close again to check for binding. If any binding is present, repeat Step 1. If there is no binding, repeat with the next throttle bore.

Step 3: After the new throttle plates are installed, open the throttle to wide open and release. Do this as fast as possible at least 10 times. Be absolutely sure that there is no binding present. Repeat all steps if necessary.

Secondary Metering Plate Installation

A secondary metering plate (used on the secondary side of 4160 carbs) has drilled restrictions rather than a metering block that has replaceable jets. Holley offers a wide range of metering plates for tuning 4160 carburetors.

Holley secondary metering plates are secured to the main body with six 8-32 x 1/2–inch flat-top screws that feature clutch-head drive; use a 5/32-inch clutch-head driver. In a pinch, you can use a small flat-blade screwdriver of the appropriate size, but it's not recommended. Clutch-head driver bits are readily available, and they're cheap (approximately $1.50 each).

A better choice is a full-handled clutch-head driver, such as NAPA Auto Parts' version (PN M-135). Clutch-head screws are used because the drive style is positive and not prone to stripping out, as long as you use a clutch-head tool.

You can replace the clutch-head screws with the "Phillips" style. If you do, be sure you use size 8-32 x 1/2–inch, in the flat-top design (flat-head face and countersink under-head). I prefer to use the style Holley uses, but some Demon carb metering plates are secured with Phillips-style screws, so either works. Sets of new clutch-head screws along with plate and bowl gaskets are available as a kit if you destroy or lose any of the original screws.

Carb-to-Intake Installation

Carburetor mounting studs have 5/16-18 threads for intake manifold engagement. Most aftermarket carb studs have 5/16-18 threads on the lower section and 5/16-24 upper threads for nut engagement. Carb studs should be hand-snugged to the intake manifold. Torque the nuts only to 60 to 80 in-lbs. Do not over-tighten the studs to the manifold.

Although you may be tempted to "double-nut" the studs to tighten them to the intake manifold, avoid doing so. Overtightening the studs to the manifold can result in slight splaying. Only apply the torque value to the nuts that engage to the studs; do not torque the studs to the 60- to 80–in-lbs value.

Once the nuts are torqued to specification, the studs are secured. This is a common mistake among novices. Never excessively tighten any stud.

Torque Values

Steel Fasteners

Size	Minimum (in-lbs)	Target (in-lbs)
No. 4	4	5
No. 5	5	7
No. 6	7	10
No. 8	12	18
No. 10	20	30
No. 12	30	50
1/4 inch	40	60
5/16 inch	90	125
3/8 inch	125	190

Fittings, Caps and Plugs

Size	Minimum (in-lbs)	Target (in-lbs)
1/16-inch NPT	50	60
1/8-inch NPT	100	150
1/4-inch NPT	160	240
5/16 inch	90	110
3/8 inch	100	130

Threaded Fasteners

Fastener	Size
Fuel-bowl screws, 4160 with side-hung, primary	12-24 x 2½ inches
Fuel-bowl screws, 4160 with side-hung, secondary	12-24 x 1⅞ inches
Fuel-bowl screws, all 4150/4500	12-24 x 2½ inches
Main jets	1/4-32
Adjustable needle/seat lock screw	1/4-32
Power valve	1/2-28
Fuel inlet to fuel-bowl threads, 4160 with banjo fitting	9/16-24
Fuel inlet to fuel-bowl threads, 4150 series	7/8-20
Fuel inlet to fuel-bowl threads, 4500 series	7/8-20
Accelerator pump mounting screws (all except 4500)	8-32 x 3/8 inch
Accelerator-pump screws, 4500 series	8-32 x 9/16 inch
Accelerator pump discharge nozzle screw	12-28 x 3/4 inch
Throttle body to main body screws	12-24 x 3/4 inch
Electric choke-housing screws	8-32 x 3/8-inch
Choke casting to main body screws	8-32 x 1⅛ inches
Mechanical secondary cam lever to throttle shaft screw	10-32 x 1/4 inch
Secondary metering plate (4160 series) clutch-head screws	8-32 x 1/2 inch
Baseplate to intake manifold, threads in intake manifold	5/16-18
Center-hung float mounting screws	6-32 x 1/2 inch
Vacuum secondary dashpot lid screws	8-32 x 1 inch

PREPARATION AND TUNING

Which intake manifold should you choose? Holley and other manufacturers offer a wide array of single- and multiple-carb manifolds. In basic terms, if you want low-end torque at lower engine RPM, consider a dual-plane intake manifold. Single-plane manifolds tend to favor upper-end power at higher engine speeds. A very broad generalization is to select a dual-plane manifold for the street and a single-plane manifold for higher-RPM operation where low-end torque isn't as critical.

Never fire a fresh engine without first pre-oiling. A pre-oiler connected to shop air forces oil through all of the oil passages, ensuring that the valvetrain and all bearings are lubed prior to starting.

Torque Requirements

Prior to mounting the carburetor on the intake manifold, you should have previously checked both the carb mounting base and the mounting surface of the manifold for flatness. Although either screws or studs may be used to mount a carburetor, the best choice by far is to use a set of carburetor mounting studs.

If you use bolts, you must be very careful not to use bolts that are too long because they can bottom out inside the blind thread holes in the manifold, preventing carb-to-manifold sealing. Also, using studs on the manifold provides an easy positioning guide for the carb.

When installing studs, install them finger-tight. Do not double-nut studs and torque them into the manifold. This can easily create a splayed condition. Tightening the nuts provides the needed clamping force.

After placing the carburetor mounting gasket in place on the intake manifold, position the carburetor onto the studs. Install flat washers and nuts finger-tight. Using an inch-pound torque wrench, tighten the nuts in a crisscross pattern to evenly distribute the clamping force. Do not overtighten the nuts, as this can easily warp the carb's mounting base.

Always refer to the carburetor's instruction sheet for recommended torque, which is typically around 60 in-lbs. Seriously, it's very easy to accidentally warp a carb base, so take the time to evenly tighten the nuts. Don't guess.

Linkage Setup

Setting up and adjusting the carburetor linkage is relatively straightforward. In the relaxed position (throttle pedal not depressed), the linkage should be adjusted so that the primary throttle shaft is not being moved or influenced by the throttle pedal.

With the engine off, manually move the carb throttle lever fully to the wide-open position and verify that the linkage doesn't prevent wide-open operation. In addition, have a helper work the throttle pedal slowly to wide open while you observe carburetor lever movement. The linkage adjustment should allow WOT, without applying additional pressure against the carb throttle lever at wide open.

Perform all of these checks with the engine off. Also, while a helper works the throttle pedal, check for smooth operation from closed to wide open. Any stiff or binding issues must be addressed.

If the carburetor has mechanical secondaries, verify that the secondary throttle plates are opening. Generally, once the primary throttle plates open approximately 40 to 45 degrees, the mechanical secondary plates should begin to open. The opening rate of the secondaries should be a bit faster than that of the primaries, with both primary and secondary plates fully open at the same time under WOT.

The most critical aspect, from a safety standpoint, is that there is no binding or sticking during any phase of throttle movement. The linkage, whether it involves rods or a cable, should connect the carb throttle lever in as straight and horizontal a position as possible. Avoid severe angles.

Engine Pre-Start Requirements

If the engine is new or freshly rebuilt, you must pre-oil it by either pressurizing the oil circuit with a pre-oil canister or removing the distributor (if so equipped) and turning the oil pump driveshaft until the oil system is fully pressurized. A mechanical oil pressure gauge should be connected (even if this is to be temporary) to verify oil pressure.

If the engine design does not use an oil pump that is shaft driven from the distributor (if the oil pump is crank driven, as on GM LS engines, for example), the only way to pressurize the oil system is by using a pressurized oil tank. Do not crank the engine to build oil pressure. An oil film should already be present at all main, rod, and cam bearings, as well as at the entire valvetrain prior to cranking the engine. Again, this applies only to an engine that is new and/or freshly rebuilt.

If the engine is equipped with a new flat-tappet camshaft rather than a roller cam, a break-in period is required for the cam and lifters. If this is the case, it's wise (but not mandatory) to first use a known, good carburetor to ensure an immediate startup and running during camshaft break-in. After the cam break-in period, you can then install the new or rebuilt carb and continue with tuning and adjustments. You don't want to find yourself fiddling with a carb problem while you're trying to seat the new cam.

You also need to make sure a few other important procedures have been performed. The fuel tank must be clean and have sufficient fuel to run the engine for a while. Be sure that the fuel offers a high-enough octane level for the particular engine. Also, be sure that the fuel system pump is in good condition and that a fresh fuel filter is installed. The vehicle should be parked in an area with good ventilation (outdoors or in a shop/garage with good ventilation). The engine's cooling system must be in good condition and the coolant level must be at specification.

Emergency Precautions

If you plan to run the engine for an extended period while tuning or adjusting the carburetor and/or engine timing, it's a good idea to place a large floor fan in front of the radiator to minimize the possibility of engine overheating. Make sure that the vehicle is parked in a safe condition (in park or neutral with the parking brake engaged). Placing wheel chocks in front of and behind the tires is a good idea.

Whenever starting an engine with a new fuel system for the first time, the concern for a fuel leak and the potential of a fire cannot be ignored. Keep a fresh, fully charged fire extinguisher, a selection of towels, and a bucket of water handy (a shop towel can smother a small fire in the case of a carburetor backfire). Think seriously about safety before starting the engine. Be prepared for a problem.

If possible, park the vehicle in such a manner that it can be pushed out of your garage in case of a fire emergency. You likely want to position the vehicle with its rear facing your bay door, so that you can open the door and allow exhaust fumes to exit. Just make sure that it is not obstructed, including by other vehicles, so that the car can be pushed out if a fire gets out of hand. If an emergency presents itself, at least you have a better chance of preventing a house fire.

Wear gloves (fire-retardant Nomex or Kevlar gloves are a good choice) and eye protection in case of a fire and in case a pressurized fuel line squirts fuel into the air. Wearing Nomex or Kevlar sleeves is also a good idea. These are available from any race supply source. They serve as arm protectors from the wrist to near the elbow. In combination with Kevlar or Nomex gloves, they protect your hands and arms from heat and sharp edges if you need to reach into a hot engine bay.

And, especially if you're working alone, be sure to keep a cell phone handy just in case you need to call 911.

Vacuum Leaks

If you hear a whistling sound and/or the idle is rough, you may have an engine vacuum leak. A vacuum leak, if left unattended, causes the engine to run lean.

Carefully retighten the carburetor mounting nuts to 60 to 80 in-lbs, in a crisscross pattern. Carefully check the intake manifold bolts for proper torque. Refer to your intake manifold instructions or factory manual. With an aluminum intake manifold, your specified torque may be 25 to 35 ft-lbs, but don't guess. Follow the specs for your application.

Check for open ports such as vacuum hose connections at the carb and manifold. Make sure that all ports are connected or plugged as required for your application. If you have trouble locating a vacuum leak, try spraying propane, carb cleaner, or WD-40 along the intake manifold-to-head mating locations and along the base of the carb. Spray slowly and along a specific route. If the engine suddenly smooths out, you've located the leak area.

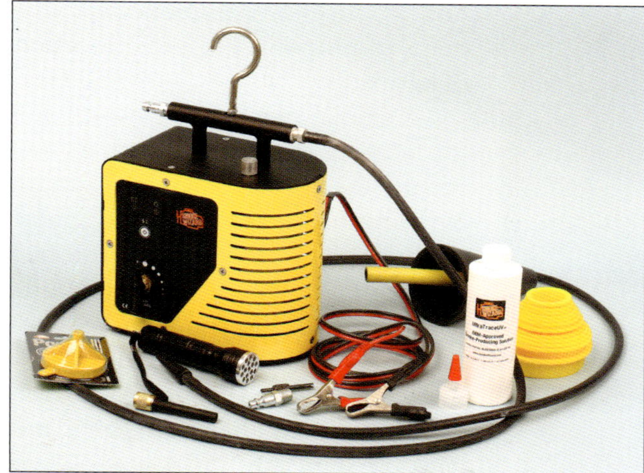

Smoke machines are made by a number of manufacturers. They are simple to use and can save lots of time by quickly locating a leak source.

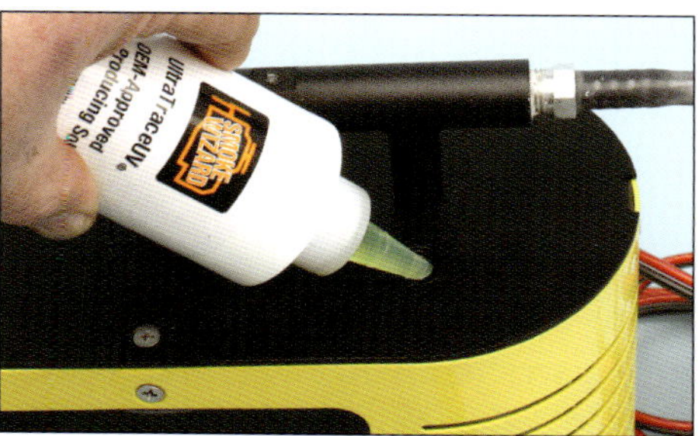

Smoke machines utilize a non-toxic and material-safe fluid, such as baby oil. The machine is connected to a 12-volt source such as the vehicle battery, along with compressed air. The fluid is gently heated and pushed into the engine at any source to check for fluid or vacuum leaks.

You can use a smoke machine to locate any number of mystery vacuum leaks.

Here, a vacuum leak was quickly found at a distributor vacuum-advance hose connection.

Another method to locate a leak (and the one I prefer) is to use a smoke machine. This is a portable unit that holds a supply of non-damaging liquid such as baby oil. The unit heats the liquid and turns it into a light smoke. A hose connects the machine to the engine. You can connect the end of the hose to a variety of locations, such as a vacuum port on the manifold. With the machine running and the engine off, look for smoke escaping from the engine. The escape location indicates the vacuum leak. You can use a UV light to help observe the smoke.

It's rare, but a vacuum leak could occur because of a crack or pinhole in a manifold casting. Vacuum leaks can also occur as the result of a cracked or cut vacuum hose or poor vacuum

When you're checking and adjusting the carburetor on the engine, never hold any small items in your hands such as screws, nuts, etc. This will help you to avoid dropping foreign objects into the carb.

hose connection. A smoke machine helps to pinpoint the exact location of a leak.

A smoke machine is also handy for locating other leaks, such as fuel or oil leaks. Fuel leaks are located by introducing smoke at any point in the fuel system, such as the fuel tank, or by disconnecting a fuel line or hose and inserting the smoke machine hose into the circuit. Allow sufficient time for the smoke to travel. The smoke won't harm anything and it won't stain any surfaces as would spraying WD-40 or carb cleaner.

Improper tightening of the intake manifold can easily result in vacuum leaks. Always use a torque wrench to tighten all intake manifold fasteners. If some bolts are difficult to access and you don't have room for a socket wrench, you can still torque to desired value using a torque wrench adapter.

Throttle Operation

Before firing the engine, carefully operate the throttle by hand to be sure that the throttle does not bind at any point, to avoid a sticking throttle, and free-revving the engine. Make sure that the throttle closes completely when fully released, and make sure that the throttle opens

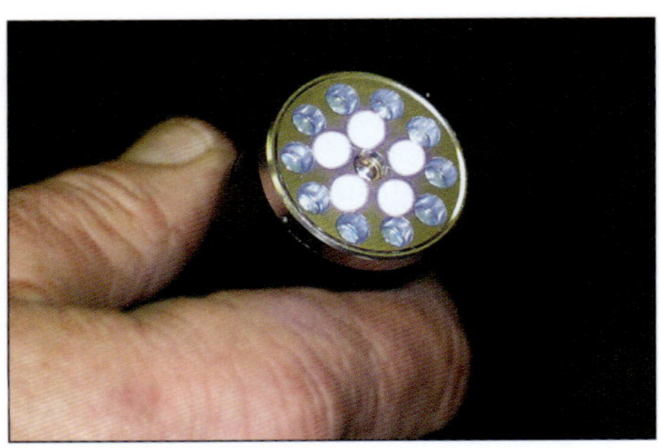

Smoke that escapes from a vacuum or fluid source is usually easy to see, but a hand-held UV light aids in spotting the leak, especially in a confined or dark area.

all the way to WOT. Don't depend only on manual checks at the carb and linkage. Have a helper work the throttle pedal while you observe throttle movement, from closed to wide open. Verify that there are no sticking or binding points, and that the throttle moves smoothly from closed to wide open. Linkage adjustments should be made now, before attempting to start the engine.

A throttle bind may be caused by an overtightened and/or unevenly tightened baseplate. Relax all four mounting fasteners that secure the carb to the manifold and retighten evenly in a crisscross pattern to the proper torque specification, which is usually 60 to 80 in-lbs. Excessive and uneven tightening of the mounting fasteners can distort the baseplate, resulting in either a vacuum leak or throttle bind, or both.

Also make sure that the carburetor mounting gasket clears the throttle plates. If the gasket has four individual holes, and the holes are too small for the throttle plates, the plates can dig into the gasket material and bind or jam the throttle plates. Depending on the intake manifold, use of a gasket with a single large square-bore opening avoids this concern.

Using a Torque Wrench

This formula obtains the correct wrench setting when using an adapter:

$$TW = L \div (L + E) \times TE$$

TW = torque wrench setting
TE = actual applied torque/desired torque
L = length of torque wrench (grip center to ratchet head)
E = length of wrench adapter extension

For example, if the torque wrench is 14 inches long and the adapter is 2 inches long, the wrench setting is the desired torque times .875 [14 ÷ (14 + 2)].

If you want 10 ft-lbs, set the wrench at 8.75 ft-lbs.
If you want 15 ft-lbs, set the wrench at 13.1 ft-lbs.
If you want 25 ft-lbs, set the wrench at 21.9 ft-lbs.
If you want 60 in-lbs, set the wrench at 52.5 in-lbs. ■

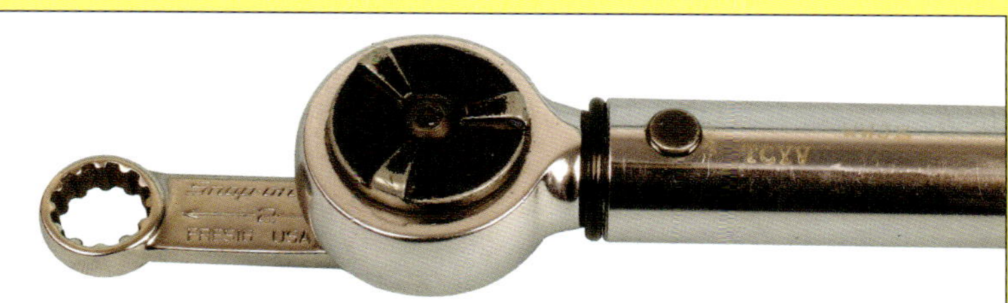

This is a 2-inch-long extension with a square female drive at one end to attach to the torque wrench and a box wrench at the opposite end to engage the bolt head. Just remember that whenever you extend the length of your torque wrench, you must re-adjust the setting on the torque wrench to compensate for the added leverage length.

Always keep the extension in-line with the torque wrench body. Pull with a smooth motion and avoid ratcheting.

Because of intake runner design, some intake manifolds have bolt locations that are difficult or impossible to access with a torque wrench and traditional socket. As a result, most people simply use an open-end or combination wrench at these locations, and guess at the torque value.

A torque wrench adapter is typically 2 inches long, center-to-center. These adapters are available in all common wrench sizes.

By adding a torque wrench adapter, you effectively lengthen the torque wrench. This added length, from the center of the grip to the wrench end, adds more leverage. Because of this, you must re-adjust the wrench setting. If you don't, you overtighten. I've never had the need to use an adapter when tightening carburetor mounting fasteners, but I've seen many intake manifold installations for which this was an absolute must.

Using a Torque Wrench *continued*

This 1/4-inch-drive inch-pound torque wrench from Mac Tools has both a visual and audible alert to let you know when you approach and achieve the desired setting. As you get close, green LED lights illuminate progressively.

As soon as you hit your desired target value, the red lights illuminate and the wrench beeps. This allows you to avoid even slight overtightening.

Carefully check throttle operation to make sure that you have no binding issues, and that the throttle is able to move freely from fully closed to WOT.

Do not overtighten carburetor mounting nuts. Most applications call for 60 to 80 in-lbs of torque when mounting a carb to an intake manifold.

Proper tightening requires the use of a torque wrench. It's too easy, especially for the novice, to overtighten and unevenly tighten, which can easily result in vacuum leaks and a warped carb baseplate.

Adjusting the operation of the accelerator pump lever to "time" pump activation in relation to the carburetor throttle movement is rather easy. With the engine off, make the adjustment at the "override" spring and screw between the pump lever and carb lever. With the throttle relaxed in the idle position, use an open-end wrench to hold the screw's locking nut (at the base of the screw) steady while turning the screw head.

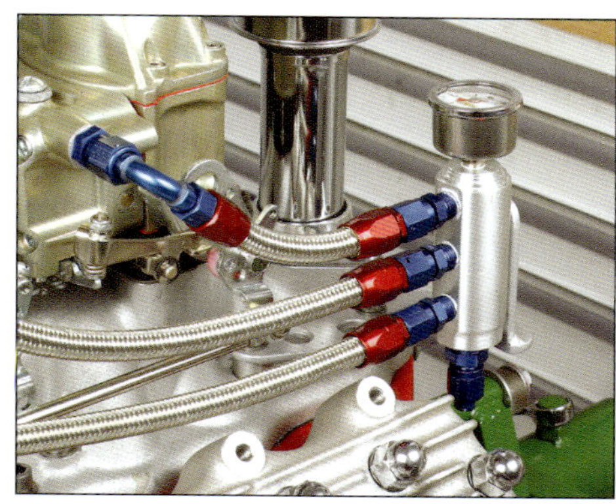

Before mounting the carburetor to the intake manifold, closely inspect the mating surfaces on both the carb baseplate and the manifold to make sure that both are clean and free of any contaminants that compromise gasket sealing.

When multiple carburetors are installed, a fuel distribution "log" allows fuel to enter from the pump and is then distributed to each carburetor. This example is on a Ford flathead, using a fabricated billet aluminum fuel log. Note the pressure gauge mounted on top of the log.

Tighten the screw, which compresses the spring, until you have freeplay between the screw and pump-lever arm. Then, slowly back off the screw, just to the point where freeplay is eliminated.

Next, move the throttle to the wide-open position. While holding at WOT, insert a feeler gauge between the pump arm and the adjuster screw. You should be able to move the pump arm an additional .015 inch. This clearance prevents the pump arm from bottoming out and placing undue stress on the arm.

Fuel Pressure

Especially if you are running an electric fuel pump, be sure to add a fuel pressure regulator. Although a fuel-injected system requires relatively high fuel pressure (depending on the application, this could be 40 to 60 psi); a carburetor requires only about 6 to 8 pounds of pressure. Running excessive fuel pressure into a carburetor overrides the needle and seat and has the potential for damaging various gaskets. Be sure to choose an electric pump that is designed for use with the lower pressure required for a carburetor. The regulator should be adjustable within a range of 6 to 8 pounds. Once the engine starts,

adjust the fuel pressure accordingly.

A fuel pressure gauge plumbed into the feed makes it easy to monitor pressure. Mounting a fuel pressure gauge can be done in a number of ways. Fuel feed tubes/bars/rails are available with 1/8-inch NPT holes, as are some fuel plumbing fittings to allow easy installation of a pressure gauge. The pressure gauge must be installed after the regulator, between the regulator and the carburetor inlet(s).

Whenever installing any NPT fitting (tapered threads) to a fuel line, the threads require a fuel-resistant sealant. Avoid using Teflon tape because threads from it can enter the fuel system and clog critical ports within the carburetor. Instead, use a Teflon pipe sealant paste. Don't

get carried away with the amount. Use only enough paste to coat the threads.

Carburetor Plumbing Techniques

Typically, a fuel feed system has 3/8- or 1/2-inch inside-diameter hose and/or tubing. When using –AN plumbing, the equivalent is size –6 or –8. The –AN hose size is easy to understand when you realize what the dash

When running nitrous injection, a separate fuel supply must be delivered to the nitrous circuit. Note the individual pressure regulators for the carb and for the nitrous system's fuel.

numbers represent. The –AN system is based on increments of 1/16 inch. A –6 size is essentially 6/16 (or 3/8) inch. A –8 size is 8/16 (or 1/2) inch. Depending on the volume of fuel required for the particular engine, a –6 size is usually adequate for street engines; a heavier-breathing performance engine may be better suited to –8 plumbing. When in doubt, going bigger to a –8 won't hurt.

Whether you opt for flexible hose or hard tubing for the majority of your plumbing, remember that you must have a flexible hose between the engine and chassis, to allow for engine rock during idle, acceleration, or deceleration.

It's always a good idea to install a fuel filter between the tank and fuel pump, regardless of whether you're using a mechanical or electric pump. However, you *must* install a filter between the fuel pump and carburetor to prevent contamination from entering the carb.

If plumbing with flexible hose, you have the option of running slip-on hose secured with clamps, or –AN hose with threaded-connection fittings and hose ends. –AN plumbing looks professional and it offers a much more secure connection. In addition, –AN hose is offered in a number of compositions (rubber lined with stainless steel exterior braiding, multi-layer heavy-duty hoses, rubber lined with black composite fabric braiding, etc.) High-performance –AN hose exterior is designed to protect the hose from abrasion.

If running hard-line tubing, common materials include aluminum and stainless steel. When custom bending is required, aluminum fuel tubing is easier to work with and is also more manageable for flaring. Hard-line tubing is easily fitted with

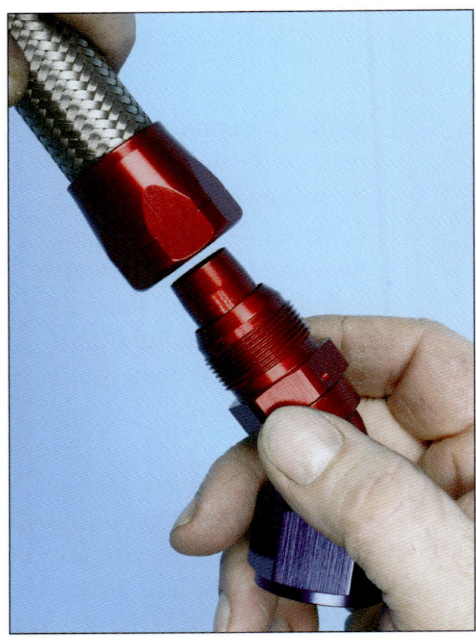

A traditional –AN hose end has a nipple tube that engages into the hose. As you thread the sections together, a compression fit secures the connection.

–AN threaded connections. The hardware attached to the tubing includes a ferrule and a tube nut.

Slip the tube nut over the tube first, followed by the ferrule, or tube sleeve. Using a flaring tool, create a single flare at the end of the tube. The tube nut is now ready to engage onto a –AN male fitting.

When plumbing from the fuel pump to the carburetor, you need to feed both fuel bowls if the carb is a double-pumper. This is most easily accomplished by using a fuel inlet feed (also called a fuel log or fuel rail, depending on design). It has a horizontal run of hose or tube, with individual fuel feeds for the primary and secondary bowls. These are available pre-assembled and ready to install to the carb. Depending on side clearance, a variety of inlets are available, from short to long, to clear any potential obstruction such as a choke mechanism.

The –AN plumbing hardware has a 37-degree seat for sealing. No additional thread-sealing compound should be necessary. Just make sure that the threads are lightly lubricated to avoid galling.

–AN plumbing provides a much more secure and robust fluid connection in contrast to a slip-on hose that's secured with a clamp. A wide range of fitting and hose-end shapes are available to suit any plumbing routing, including but not limited to straight, 45-degree (shown), 90-degree, and 180 degree.

Although a fuel injection system requires relatively high fuel pressure, perhaps in the 40- to 50-psi range, a carburetor requires only about 6 to 8 psi. If fuel pressure is excessive, it overrides the needle and seat and floods the carb.

If you're running an electric fuel pump, be sure to obtain a pump that's rated for use with a carb. Because the pump may be able to provide higher pressure, you should always install an adjustable fuel pressure regulator between the pump and carb. Even if you're running a mechanical fuel pump, using a pressure regulator is still a good idea. In conjunction with an adjustable regulator, you should install a fuel pressure gauge, which is located between the regulator and carb.

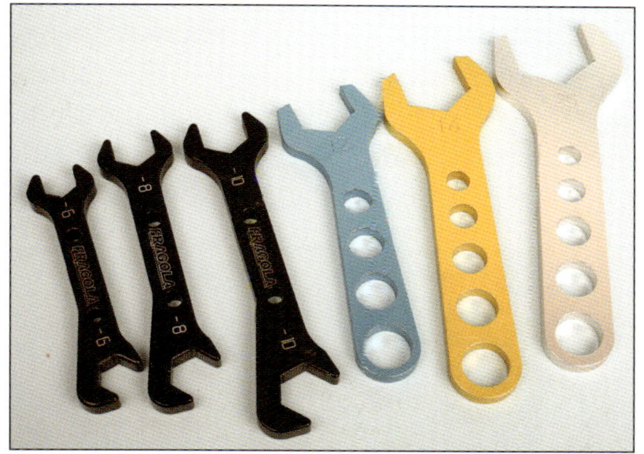

Specialty aluminum wrenches for the range of –AN sizes are readily available from several sources. If you have –AN aluminum plumbing on your vehicle, you need a set of these tools.

Whenever you deal with aluminum –AN fittings and hose ends, always use an aluminum wrench to avoid damaging the surface.

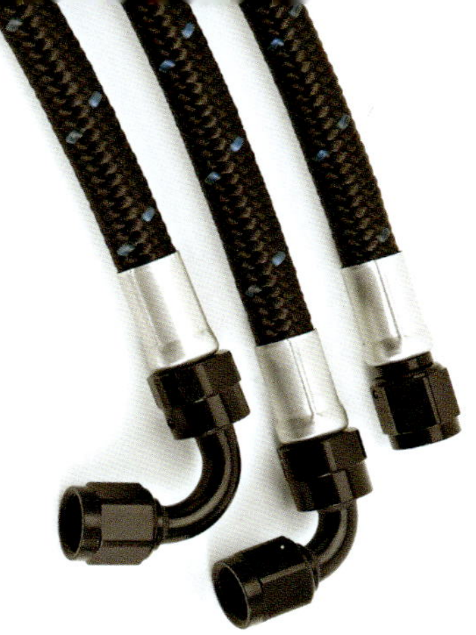

An option to traditional –AN hose-to-hose end construction is the crimp-collar construction, which provides a secure connection with reduced weight. A special crimping machine is required to assemble this style.

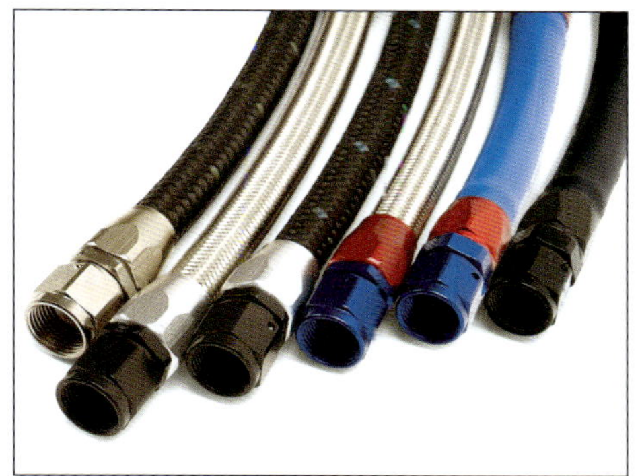

Fuel-compatible high-performance hose is available in a variety of construction, including rubber-lined stainless braid, high-temperature synthetic braid, and rubber/Kevlar.

If a fuel line runs close to a high-heat source, thermal shielding is available to insulate the fuel hose. This tough shielding also provides additional protection from abrasion.

A common method of feeding a double-pumper carb is with a fuel feed created from –AN stand-offs, –AN fittings, and flexible braided fuel hose. You may assemble the fuel feed yourself or purchase it ready-to-install, based on the center-to-center fuel inlet distance of the specific carburetor model.

Fuel inlet fittings that accept –AN hose ends are available in short "stock" length.

Extended-inlet fittings are available in a range of lengths.

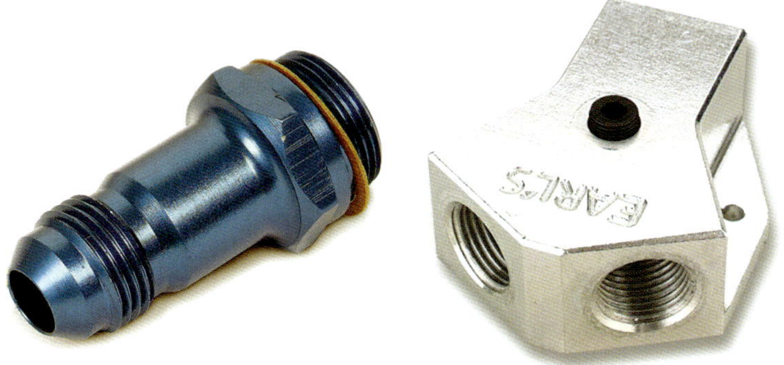

A simple way to feed both bowls of a Double Pumper carb is with a three-way distribution block.

Many carburetor fuel inlet logs have 1/8-inch-NPT-threaded ports to allow the installation of a pressure gauge. With a gauge in place, you can adjust and verify your fuel pressure.

When selecting a pressure gauge, the better choice is a liquid-filled gauge. It dampens the gauge needle to allow more accurate observation of fuel pressure, compensating for engine vibration. A non-dampened gauge, especially on an engine with a radical camshaft, allows the gauge needle to bounce erratically, making it difficult to pinpoint the adjusted pressure.

Don't forget to add a fuel filter. A filter should be installed between the pump and carb or between the pump and fuel pressure regulator.

Extended fuel inlets, often called stand-offs, make it easy to plumb the carb if clearance issues are present, such as the need to clear a choke mechanism.

This overhead view illustrates the need for longer inlet fittings.

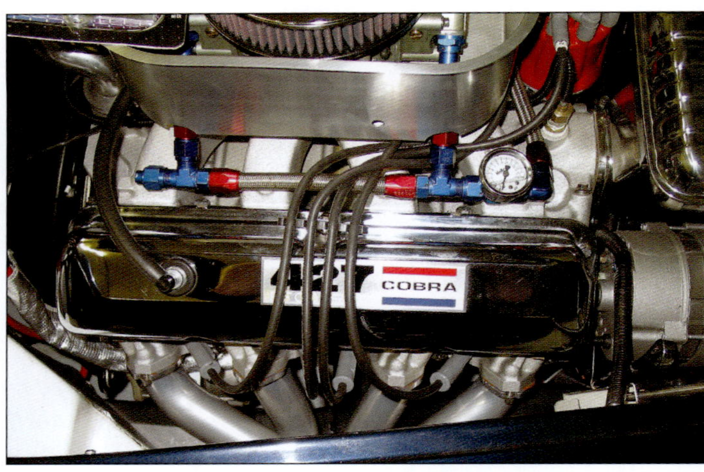

Note the fuel pressure gauge on this fuel rail. The opposite end of the rail is sealed with a –AN cap.

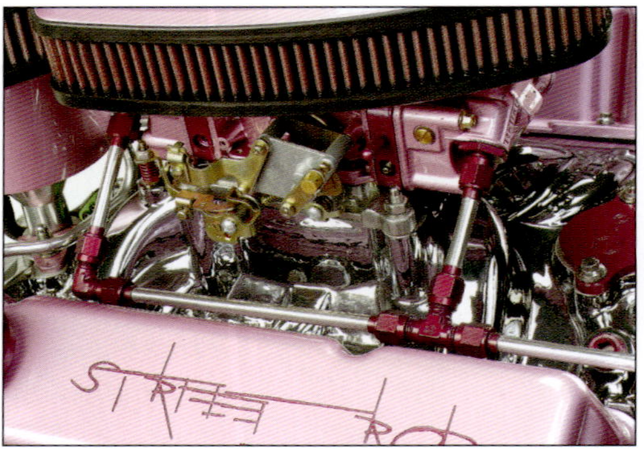

In this example, the carb was plumbed with aluminum fuel line tubing, assembled with –6 tube sleeves and tube nuts.

Although a slip-on hose and clamp suffices, plumbing with –AN hardware provides a more secure connection.

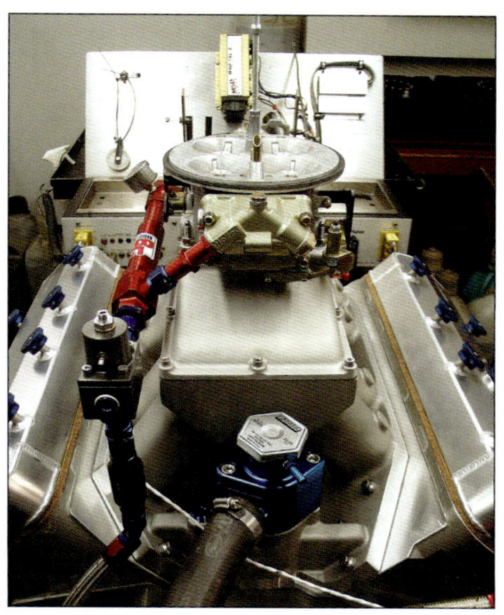

This fuel inlet plumbing has an adjustable fuel-pressure regulator mounted before the fuel rail, with a fuel pressure gauge mounted on the fuel rail. The gauge can be mounted anywhere along the fuel inlet rail, as long as it's between the regulator and the carb.

Carburetor fuel inlet fittings are available in several styles. If the hex is very close to the carb body, you may not be able to use a line wrench. If using an open-end wrench, take your time and make sure that the wrench is fully and firmly seated on the fitting to avoid slipping and damaging the fitting.

Note the extended fuel feed stand-offs on this fuel rail. Various lengths of fuel inlet fittings/extensions are available, allowing you to mount the fuel rail as far away from the carb as necessary to clear any obstructions. Manufacturers including Aeroquip, Fragola, Russell, Earls, Goodridge, XRP, and others offer a wide range of carburetor fuel inlet fittings to suit any application.

Spacers made of a composite or phenolic material can help to reduce heat-soak at the carburetor. If a spacer is used, longer carburetor mounting studs are required.

Note the milled air reliefs on this carb spacer. They serve to further aid in heat dissipation.

An aluminum carburetor spacer not only slightly extends the air column but also serves to dissipate heat. When a spacer is used, a gasket is required between the manifold flange and spacer as well as between the spacer and carb.

Linkage rods are available with radius rod ends for friction-free pivoting. Rods should have right-hand threads at one end and left-hand threads at the other end, to allow fine-tuning of length adjustment without the need to disconnect one end. When the length has been properly adjusted, be sure to tighten the jam nut at each end to prevent the rod from loosening and changing length.

Dual Carbs

When running twin carburetors, a distribution block or fuel manifold is required to feed both carbs from a single primary feed line. In addition, throttle linkage must be carefully set up and adjusted to allow each carburetor's throttle plates to operate in unison or progressively as needed. Linkage rods featuring threaded rods equipped with radiused rod ends provide reduced friction and smooth operation.

Whether the carburetors are mounted in-line or side-to-side (as with a cross-ram setup) is often dictated by the configuration of the intake manifold and available clearance. The operating rods should be oriented in as horizontal a manner as possible. Avoid severe angles that can alter lever ratios and that can potentially cause a binding situation. The throttle linkage base must be firmly anchored, usually to the intake manifold.

The vertical rod that connects the base to the carb rods should be as vertical as possible. Plan on spending a bit of time to design the linkage rod system. The objective is to attain smooth operation, with full throttle closure at rest and full WOT without any binding or excess friction.

Dyno Testing Your Setup

Not everyone has access to an engine dyno, but running the engine on a test cell and gathering data can help to adjust the carburetor and tune it for maximum power before installing the engine. Bear in mind that as you tune a carb (and engine timing as well) for maximum horsepower and torque, power and torque are developed at various engine speeds. The dyno gets you into the ballpark. Once the engine is installed to the vehicle, further tweaking may be required to compensate for vehicle weight, tire diameter, operational speeds and loads, etc. Gathering engine dyno data allows you to know what the engine is capable of and provides you with a quantitative reference.

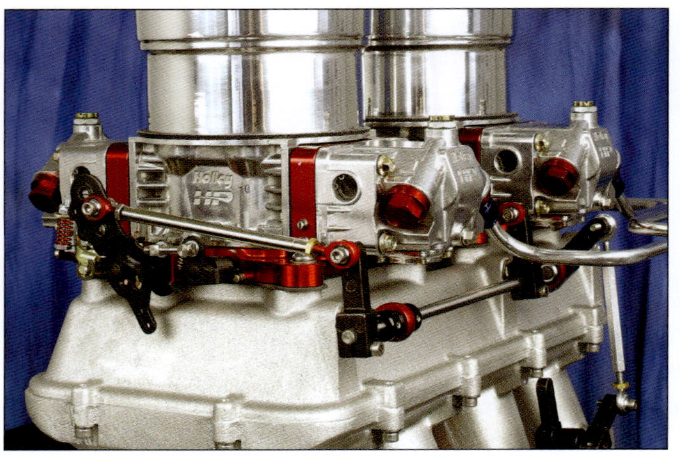

Setting up a dual-carb linkage on a sideways-mounted carb requires careful attention to rod length and length adjustment. Radius rod ends provide low-friction for all pivot points. A primary horizontal shaft connects the actuator rods to each carb.

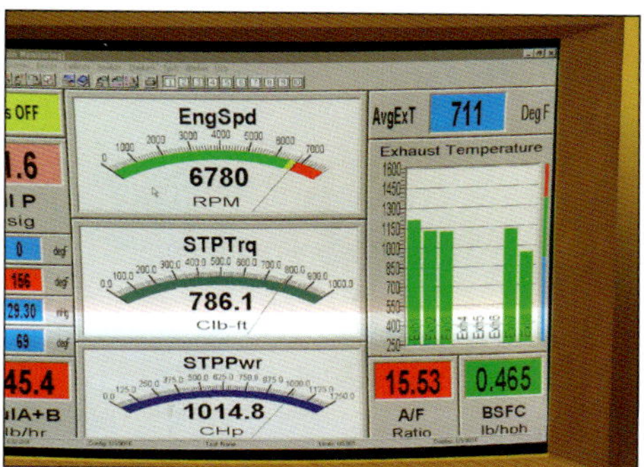

After installing the carb, you inevitably have to spend some time calibrating it for a particular engine and set of conditions. Always have your tools handy before starting the engine. You may need to adjust the idle stop, idle-mixture screws, float level, and/or re-tighten manifold or carb mounting fasteners.

Although vehicle weight, rear axle ratio, and tire diameter all factor into final engine tuning, being able to quantify the engine prior to vehicle installation is invaluable. Rather than guessing at horsepower and torque, at what RPM points these occur, an engine dyno provides this information and allows you tune and adjust the fuel and spark, providing an informed baseline. This particular engine pulled 1,014 hp initially, and managed to pull 1,115 hp and almost 900 ft-lbs of torque, after a wee bit of carb tuning and ignition curve adjustments.

An engine dyno allows you to fully control and monitor the engine while verifying operation and testing/ tuning for maximum horsepower and torque. Remember that an engine dyno provides data only on the engine. Additional tuning and tweaking may be needed after the engine is installed into a vehicle. A chassis dyno allows you to monitor and tune while gathering both engine data and power applied to the drive wheels.

Pre-Oiling a Fresh Engine

It may seem out of place to discuss engine pre-oiling in a carburetor book, but it's extremely important to note that if you are dealing with a new engine that's never been run before, or an engine that's been sitting in storage for a long period of time, it's critical to make sure that the bearings and valvetrain are properly lubricated prior to firing the engine for the first time.

Do not disconnect the coil(s) and crank the engine to establish oil pressure. The proper method is to prime the oiling circuit without trying to crank the engine via the starter. This can be done in several ways, depending on the type of engine. If the engine has a distributor and an oil pump that's driven by the distributor shaft-to-oil pump shaft, you can remove the distributor and drive the oil pump with a heavy-duty electric drill or pneumatic drill, using a drive adapter that engages the oil pump input shaft.

If the engine has a crankshaft-driven oil pump (this refers to many late-model engine designs, such as GM's LS), you can't run the pump with a drill. In this case, using a pressurized engine pre-oiler canister is the answer. Actually, this type of delivery system works for any type of engine.

The canister holds engine oil and is pressurized with compressed air. A hose connects the canister to the engine block, at an oil entry port. Once the canister's valve is opened, oil is pushed into the engine, delivering oil to all bearings and the valvetrain. Regardless of the approach you use, remove the valvecovers to expose the rockers. Pressurize the oiling system until you see oil exiting at all of the rocker locations. In some cases, this may take a few minutes.

It's a good idea to manually rotate the crankshaft slowly during the oil priming process so that at some point the oil feed holes in the bearings align directly to the oil passages in the main saddles and in some camshafts. During oil pressurization, slowly rotate the crank two full turns. The crank rotation may not be absolutely necessary, but it's a good idea.

Setting and adjusting the engine's ignition timing goes hand in hand with carburetor adjustment when pursuing optimum idle and maximum power.

Prior to firing the engine, double-check all linkage attachments to prevent linkage, screws, or nuts from vibrating off.

velocity stack can effectively lengthen the flow path of the intake manifold runners on a manifold with isolated runners (rather than a dual-plane manifold), potentially improving cylinder filling. Basically, a velocity stack can help to tune the length of the air column that enters the engine. A velocity stack, because of its vertical air intake design, needs sufficient space above the top bell of the stack, so the application should have no hood or if a hood is on the vehicle, you should maintain at least 2 inches of space between the hood and the stack.

Many variables come into play, but using velocity stacks has the potential to improve airflow and power, although the improvement may be minimal. The only reason to run a velocity stack on a street engine is for the looks. If you desire a velocity stack for a V-8 engine, a good source is a performance marine distributor because they are popular for high-performance boat applications.

Float Height

Place a rag under the fuel bowl to catch any potential fuel spill. If the fuel bowl is equipped with a brass sight plug, remove the plug and its gasket. With the vehicle on a level surface and with the engine idling, hold the needle and seat float-adjustment hex steady with a 5/8-inch box wrench and loosen the lock screw with a broad flat-blade screwdriver. Then, rotate the hex slowly to adjust float level. Turn the hex counter-clockwise to raise the fuel level or clockwise to lower it until the fuel level is at the bottom of the sight hole. Take your time and adjust it in small, slow increments; wait a few seconds until the fuel level stabilizes.

When you know that oil has been distributed throughout the oil circuit, you can then turn your attention to firing the engine and checking and adjusting the carburetor.

Engine pre-oiler canisters are available from various sources. Some are steel (basically the same construction as a propane tank); others are made of aluminum. The aluminum units prevent potential rust scale buildup inside the tank that might be present in an improperly stored steel unit.

Air Cleaners

Unless you're running the engine in a dust-free environment (such as marine use), you need a filter that prevents contaminants from entering the engine. With that said, pay attention to the air cleaner base (the bottom section). Make sure that you retain at least 3/8-inch clearance between the air cleaner top and the carburetor's vent tubes. This is necessary whether the vent tubes are square cut or angle cut. If the base is too close to the vent tubes, it can mess up the reference pressures for the carburetor's bowls.

An air cleaner provides more than simply filtering out airborne debris. An air cleaner with a filter element also protects against potential fires caused by backfiring.

Velocity stacks are sometimes chosen simply because they look cool. Regardless of the appearance factor, a

When installing an air cleaner lid, don't get carried away. If you use excessive force, you might place undue stress on the carburetor's main body casting.

Individual velocity stacks provide isolated airflow paths for each carburetor. The length of the stack extends the airflow column of the intake manifold runners.

Hold the hex steady with the 5/8-inch wrench and tighten the lock screw. Rev the engine a few times and then allow it to return to idle to verify that fuel level hasn't changed. Reinstall the brass sight plug and its gasket. Some Holley carbs have a fixed-glass sight window that allows you to observe the fuel level without having to remove a sight plug.

Idle Mixture

Initially, turn the idle-mixture screws in fully and gently (clockwise), until the screw stops, then back it out by one turn. Connect a vacuum gauge to the engine at a manifold vacuum source. With the engine running, adjust the mixture screw on one side of the metering block until you achieve the highest vacuum reading. Go to the opposite side and repeat the process. Then go back to the first side and repeat, then to the opposite side and repeat again. Do this until you have achieved the highest stable vacuum reading.

If the carb is equipped with idle-mixture screws on the secondary metering block, perform the same procedure.

Generally speaking, an acceptable idle mixture should be obtained with the mixture screws adjusted to somewhere between 1/2 and 2 turns out from fully closed.

Velocity stacks are available in several heights to accommodate Holley 4160-, 4150-, and 4500-series carburetors. Unless your application is a marine engine, be sure to use screen or mesh filters at the base of each stack.

Once all of the mixture screws are adjusted, re-adjust the idle speed to the desired RPM. You may need to re-check idle-mixture adjustments. The engine should be warmed to normal operating temperature to make sure that the fast-idle

This custom LS, bored and stroked to 408 ci, was outfitted with dual Holley 600-cfm Ultra carbs, and pulled almost 700 hp on dyno.

When adjusting needle and seat height, hold the nut stationary with a 5/8-inch wrench while loosening the locking screw. Then turn the wrench to raise or lower the needle and seat. Once adjusted, hold the nut steady and tighten the locking screws.

When initially adjusting the idle-mixture screws on each side of the metering block, begin by turning the screw all the way in until it gently bottoms out, then turn the screw counterclockwise 1½ turns. This provides a starting point for further adjustment as the engine runs.

cam was released before final adjustments.

A needle screw on each side of the primary metering block adjusts idle mixture. This is called a two-corner idle. Some carbs may have idle screws on the secondary metering block as well; this is known as a four-corner idle system.

On "non-emissions" carbs, turn the screw clockwise to lean the mixture and counterclockwise to richen the mixture. As the screw is turned inward (clockwise), the air/fuel flow to the curb-idle discharge port is reduced, leaning the mixture. If the screw is seated fully inward, this flow is stopped.

With the engine off, turn each idle-mixture screw inward (clockwise) until the screw lightly bottoms out. Then turn the screw counterclockwise by 1½ turns. Do this to all idle-mixture screws. This provides a starting point.

Connect a vacuum gauge to a full manifold vacuum port. If the vehicle is equipped with a manual transmission, leave the transmission in neutral with the parking brake on.

If equipped with an automatic transmission, set the parking brake and, once the engine has started, place the transmission in Park. Start the engine and allow it to warm up to operating temperature.

Adjust each mixture screw to achieve the highest vacuum reading. Adjust each screw by the same amount.

Once the idle mixture has been adjusted, you may then adjust curb-idle speed by turning the idle-speed adjuster screw on the carb linkage. Once idle speed has been adjusted, re-check all mixture screws, again with the goal of achieving the highest vacuum reading.

Engine Bogs

If the engine "bogs" upon acceleration, chances are that the accelerator pump shot isn't sufficient. Curing this may involve moving to a larger discharge nozzle, a larger capacity accelerator pump, or both. Changing to a camshaft with more duration can require a greater pump shot.

Depending on other variables, such as vehicle weight and stall speed of an automatic transmission torque converter, moving to a smaller discharge nozzle may help the existing accelerator pump do its job.

Again, always make only one change at a time, otherwise you won't know which change helped or hindered performance. As engine load increases (using a lighter flywheel, moving to a lower-RPM torque converter, etc.), you may be able to reduce discharge nozzle size. As the engine load increases (moving to a higher stall converter or a heavier flywheel), you may have to increase discharge nozzle size, and possibly go to a larger accelerator pump.

Replacing Jets

f you feel the need to alter your Holley carb, stick to the rule of making only one change at a time, whether that involves a jet change, power-valve change, accelerator pump change, idle-mixture adjustment, etc. If you make more than one alteration at a time, you won't know which change improved or degraded performance.

With regard to jet changes: Just replace the jets. Don't drill out the existing jets. Holley offers an extensive array of jet sizes to suit any application, and they're not very expensive, so it just doesn't make sense to increase jet size by drilling. Besides, drilling can create out-of-round holes and burrs that alter the desired orifice size and flow.

If you intend to experiment with jet changes, organize the spare jets on a clean jet card or in a clean compartmentalized storage tray. Don't string a bunch of jets together with a piece of wire; this can result in damaging the orifices.

Moreover, when making any changes, keep a record of your changes instead of relying on your memory. If you swapped out size 72 jets for a pair of 74 jets, and a couple of weeks later want to make another jet change, you might not remember what size jet was initially removed. Record every change that you make so that you have a basis of comparison for future reference. And be sure to date each change. ∎

Experiment with discharge nozzle size. When you obtain a crisp throttle response, record the nozzle size, then move up an additional .002 inch and test the response. Chances are that you are in the right ballpark.

Check clearance between the accelerator pump lever and the adjuster screw. You should have about .015 to .020 inch of clearance between the pump lever and the adjuster when the throttle is moved to WOT. Check the clearance between the accelerator pump lever and the pump lever cam. If you lowered the idle speed, it moves the cam a bit farther away from the lever, which delays the pump shot. Adjust the plastic cam so that there is zero clearance between the pump actuating lever and the cam.

Power Valve

You need a vacuum gauge to monitor intake manifold vacuum and to determine when the power valve opens. The correct power valve for the application should provide a slightly higher opening point than the highest manifold vacuum to protect the engine from a lean condition. However, if manifold vacuum is very low at idle or part-throttle, a lower opening point may be needed to obtain a power valve that doesn't open until the engine needs the added fuel.

For example, if the carb is equipped with a 65 power valve, and the engine vacuum sometimes drops to, say, 5 Hg at idle, installing a 40 power valve that opens at 4.0 Hg may help. For WOT use, such as racing, always use a power valve that opens slightly higher than the highest manifold vacuum to avoid a lean condition.

You need to be aware of the relationship between main jets and the power valve. All too often, "tuning" for performance is done by simply changing jets to maximize power at WOT. If jet size is increased and you're not happy with WOT power, you may be allowing too much fuel to enter at part throttle, even when the power valve circuit is closed at high manifold vacuum. You're better off tuning the power-valve circuit first (by changing the power valve) to maximize efficiency at part throttle, and then changing jets to optimize power at WOT.

Tuning Vacuum Secondary Operation

Vacuum-operated secondary systems are intended to open the secondaries when the engine demands additional fuel and air. Ideally the transition should be smooth and unnoticed by the driver. Drivers may think that the secondaries aren't working properly if they can't "feel" an abrupt kick in the pants. If a sudden surge is felt, it indicates that there was a split-second transition when the engine stumbled and "hiccupped" before surging forward. The opening rate of the vacuum secondary system can be tailored by changing the spring inside the vacuum housing.

Holley offers a kit (PN 20-13) that contains several color-coded springs. If the vehicle is light and the engine is large, a lighter spring can be used to accommodate a faster engine speed increase. Conversely, a heavier vehicle with a smaller engine can benefit from using a stiffer spring to avoid a stumble upon acceleration. Just remember that a vacuum secondary system, when tuned properly, should provide a seamless transition when the secondaries open.

Carb Number	Carb Model Number	CFM	Renew Kit	Trick Kit	Needle & Seat	Primary Main Jet	Secondary Main Jet or plate	Primary Metering Block	Secondary Metering Block	Power Valve	Pump Nozzle Size, Primary	Secondary Nozzle Size or Spring Color
R1848-1	4160	465	37-119	37-933	6-506	122-57	N/S	N/S	34R9716-3	125-85	0.025	Green
R1849	4160	550	37-119	37-933	6-506	122-62	N/S	N/S	N/S	125-85	0.025	Plain
R1850-2	4160	600	37-119	37-933	6-506	122-66	134-9	134-128	134-9	125-65	0.025	Plain
R1850-3	4160	600	37-119	37-933	6-506	122-66	134-9	134-128	134-9	125-65	0.025	Plain
R1850-4	4160	600	37-119	37-933	6-506	122-66	134-9	134-128	134-9	125-65	0.031	Plain
R1850-5	4160	600	37-119	37-933	6-506	122-66	134-9	134-128S	134-9	125-65	0.031	Plain
R1850-6	4160	600	37-119	37-933	6-506	122-66	134-9	134-128	134-9	125-65	0.031	Plain
R1850-7	4160	600	37-119	37-933	6-506	122-64	134-9	N/A	134-9	125-65	0.031	Plain
R1850-8	4160	600	37-119	37-933	6-506	122-64	134-9	N/A	134-9	125-65	0.031	Plain
R1850-9	4160	600	37-119	37-933	6-511	122-64	134-9	N/A	134-9	125-65	0.031	Plain
R1850-10	4160	600	37-119	37-933	6-511	122-64	134-9	N/A	134-9	125-65	0.031	Plain
R2818-1	4150	600	37-1537	37-933	6-506	122-65	122-76	N/A	N/S	125-65	0.025	Purple
R3124	4150	750	37-1539	37-933	6-504	122-70	122-76	N/S	N/S	125-85 (12)	0.025	Yellow
R3247	4150	780	37-1539	37-933	6-504	122-70	122-76	N/S	N/S	125-85 (12)	0.021	Yellow
R3259-1	4150	725	N/A	N/A	N/S	122-68	122-78	N/S	N/S	125-85	0.025	Yellow
R3310-1	4150	780	37-1539	37-933	6-504	122-72	122-76	134-131	N/S	"(12,13)"	0.025	Plain
R3310-2	4160	750	37-754	37-933	6-504	122-72	134-21	134-131	134-21	125-65	0.025	Plain
R3310-3	4160	750	37-754	37-933	6-504	122-72	134-21	134-131	134-21	125-65	0.025	Plain
R3310-4	4160	750	37-754	37-933	6-504	122-72	134-21	134-131	134-21	125-65	0.031	Plain
R3310-5	4160	750	37-754	37-933	6-504	122-72	134-21	134-131	134-21	125-65	0.031	Plain
R3310-6	4160	750	37-754	37-933	6-504	122-72	134-21	134-131	134-21	125-65	0.031	Plain
R3310-7	4160	750	37-754	37-933	6-504	122-72	134-21	134-131	134-21	125-65	0.031	Plain
R3310-8	4160	750	37-754	37-933	6-504	122-72	134-21	134-131	134-21	125-65	0.031	Plain
R3310-9	4160	750	37-754	37-933	6-504	122-72	134-21	134-131	134-21	125-65	0.031	Plain
R3310-10	4160	750	37-754	37-933	6-504	122-70	134-21	N/A	134-21	125-65	0.031	Plain
R3310-11	4160	750	37-754	37-933	6-504	122-70	134-21	N/A	134-21	125-65	0.031	Plain
R3367	4160	585	37-119	37-933	N/S	122-65	34R9716-22	N/S	N/R	125-65	0.025	Purple
R3370	4160	585	37-119	37-933	6-504	122-65	N/S	N/S	N/R	125-65	0.025	Purple
R3418-1	4150	855	37-1539	37-933	6-504	78C/82T	82C/80T	N/S	N/S	(15,21)	0.028	Yellow
R3613	4150	770	37-1539	37-933	6-504	122-71	122-76	N/S	N/S	125-85 (12)	0.021	Yellow
R3659	2300	466	37-1537	37-933	6-504	N/R	N/S	N/R	N/S	N/R	N/R	Brown
R3660	2300	350	37-1537	37-933	6-504	122-64	N/R	N/S	N/R	125-65	0.021	N/R
R3807	4150	595	37-1537	37-933	N/S	122-67	122-72	N/S	N/S	125-65	0.025	Purple
R3810	4160	585	37-1537	37-933	N/S	122-65	34R9716-22	N/S	N/R	125-65	0.025	Purple
R3811	4160	585	37-1537	37-933	N/S	122-65	N/S	N/S	N/R	125-65	0.025	Purple
R3910	4150	780	37-1539	37-933	6-504	122-71	122-76	N/S	N/S	125-65 (12)	0.021	Yellow
R4053	4150	780	37-1539	37-933	6-504	122-68	122-76	N/S	N/S	125-65 (12)	0.025	Yellow
R4055-1	2300	350	37-1537	37-933	6-504	122-63	N/R	N/S	N/R	125-65	0.021	N/R
R4056-1	2300	350	37-1537	37-933	6-504	122-61	N/R	N/S	N/R	125-65	0.025	N/R
R4118	4150	725	37-1539	37-933	6-504	122-68	122-78	N/S	N/S	125-85	0.025	Yellow
R4144-1	2300	350	37-1537	37-933	6-504	122-62	N/R	N/S	N/R	125-65	0.031	N/R
R4224	4160	660	37-1537	37-933	6-508	122-76	34R9716-12	N/A	34R9716-12	N/R	0.025	0.025
R4235	4160	770	N/A	37-933	6-504	75/80	N/S	N/S	N/R	125-65	0.035	Plain
R4236	4160	770	N/A	37-933	6-504	122-80	N/S	N/S	N/R	125-65	0.035	Plain
R4295	4150	585	37-485	37-933	6-504	122-69	122-71	N/S	N/S	125-65	0.025	0.025
R4296	4150	850	37-485	37-933	6-504	78C/82T	82C/80T	N/S	N/S	125-65 (15)	0.035	0.035
R4346	4150	780	37-1539	37-933	6-504	122-68	122-76	N/S	N/S	125-85 (12)	0.025	Yellow
R4365-1	2300	500	37-1537	37-933	6-504	N/R	N/S	N/R	N/R	N/R	N/R	Yellow

Primary Bowl Gasket	Primary Metering Block Gasket	Secondary Bowl Gasket	Secondary Metering Block Gasket	Secondary Metering Plate Gasket	Primary Fuel Bowl	Secondary Fuel Bowl	Throttle Body & Shaft Assembly	Venturii Diameter Primary	Venturii Diameter Secondary	Throttle Bore Diameter Primary	Throttle Bore Diameter Secondary
108-83-2	108-89-2	108-90-2	108-90-2	108-27-2	N/A	134-105	N/S	1-3/32	1-3/32	1-1/2	1-1/2
108-83-2	108-89-2	108-90-2	108-90-2	108-27-2	N/A	134-105	N/S	1-3/16	1-1/4	1-1/2	1-1/2
108-83-2	108-89-2	108-90-2	108-90-2	108-27-2	134-101	134-105	112-20	1-1/4	1-5/16	1-9/16	1-9/16
108-83-2	108-89-2	108-90-2	108-90-2	N/R	134-101	134-105	112-20	1-1/4	1-5/16	1-9/16	1-9/16
108-83-2	108-89-2	108-90-2	108-90-2	N/R	134-101	134-105	112-20	1-1/4	1-5/16	1-9/16	1-9/16
108-83-2	108-89-2	108-90-2	108-90-2	N/R	134-101S	134-105S	112-20	1-1/4	1-5/16	1-9/16	1-9/16
108-83-2	108-89-2	108-90-2	108-90-2	N/R	134-101	134-105	112-20	1-1/4	1-5/16	1-9/16	1-9/16
108-83-2	108-89-2	108-90-2	108-90-2	N/R	134-101S	134-105S	112-20	1-1/4	1-5/16	1-9/16	1-9/16
108-83-2	108-89-2	108-90-2	108-90-2	N/R	134-101	134-105	112-20	1-1/4	1-5/16	1-9/16	1-9/16
108-83-2	108-89-2	108-90-2	108-90-2	N/R	134-281S	134-282S	112-20	1-1/4	1-5/16	1-9/16	1-9/16
108-83-2	108-89-2	108-90-2	108-90-2	N/R	134-281S	134-282S	112-112	1-1/4	1-5/16	1-9/16	1-9/16
108-83-2	108-89-2	108-83-2	108-89-2	N/R	134-101	N/A	N/S	1-1/4	1-5/16	1-9/16	1-9/16
108-83-2	108-89-2	108-83-2	108-89-2	N/R	N/S	N/S	N/S	1-3/8	1-7/16	1-11/16	1-11/16
108-83-2	108-89-2	108-83-2	108-89-2	N/R	N/S	134-102	N/S	1-3/8	1-7/16	1-11/16	1-11/16
108-83-2	108-89-2	108-83-2	108-89-2	N/R	N/S	N/S	N/S	1-5/16	1-3/8	1-11/16	1-11/16
108-83-2	108-89-2	108-83-2	108-89-2	N/R	134-103	134-102	112-9	1-3/8	1-7/16	1-11/16	1-11/16
108-83-2	108-89-2	108-90-2	108-90-2	108-27-2	134-103	134-102	112-9	1-3/8	1-7/16	1-11/16	1-11/16
108-83-2	108-89-2	108-90-2	108-90-2	108-27-2	134-103	134-102	112-9	1-3/8	1-7/16	1-11/16	1-11/16
108-83-2	108-89-2	108-90-2	108-90-2	108-27-2	134-103	134-102	112-9	1-3/8	1-7/16	1-11/16	1-11/16
108-83-2	108-89-2	108-90-2	108-90-2	108-27-2	134-103S	134-102S	N/A	1-3/8	1-7/16	1-11/16	1-11/16
108-83-2	108-89-2	108-90-2	108-90-2	108-27-2	134-103S	134-102S	N/A	1-3/8	1-7/16	1-11/16	1-11/16
108-83-2	108-89-2	108-90-2	108-90-2	108-27-2	134-103	134-102	N/A	1-3/8	1-7/16	1-11/16	1-11/16
108-83-2	108-89-2	108-90-2	108-90-2	108-27-2	134-103	134-102	N/A	1-3/8	1-7/16	1-11/16	1-11/16
108-83-2	108-89-2	108-90-2	108-90-2	108-27-2	134-103	134-102	N/A	1-3/8	1-7/16	1-11/16	1-11/16
108-83-2	108-89-2	108-90-2	108-90-2	108-27-2	134-103	134-102	112-117	1-3/8	1-7/16	1-11/16	1-11/16
108-83-2	108-89-2	108-90-2	108-90-2	108-27-2	N/S	134-105	N/S	1-1/4	1-5/16	1-9/16	1-9/16
108-83-2	108-89-2	108-90-2	108-90-2	108-27-2	N/S	134-105	N/S	1-1/4	1-5/16	1-9/16	1-9/16
108-83-2	108-89-2	108-83-2	108-89-2	N/R	134-103	N/S	N/S	1-9/16	1-9/16	1-3/4	1-3/4
108-83-2	108-89-2	108-83-2	108-89-2	N/R	N/S	N/S	N/S	1-3/8	1-7/16	1-11/16	1-11/16
N/R	N/R	108-90-2	108-90-2	108-27-2	N/R	N/S	N/S	1-3/8	N/A	1-3/4	N/A
108-83-2	108-89-2	N/R	N/R	N/R	N/S	N/R	N/S	1-3/16	N/A	1-1/2	N/A
108-83-2	108-89-2	108-83-2	108-89-2	N/R	N/S	N/S	N/S	1-1/4	1-5/16	1-9/16	1-9/16
108-83-2	108-89-2	108-90-2	108-90-2	108-27-2	N/S	N/S	N/S	1-1/4	1-5/16	1-9/16	1-9/16
108-83-2	108-89-2	108-90-2	108-90-2	108-27-2	N/S	N/S	N/S	1-1/4	1-5/16	1-9/16	1-9/16
108-83-2	108-89-2	108-83-2	108-89-2	N/R	134-103	N/S	N/S	1-3/8	1-7/16	1-11/16	1-11/16
108-83-2	108-89-2	108-83-2	108-89-2	N/R	134-103	N/S	N/S	1-3/8	1-7/16	1-11/16	1-11/16
108-83-2	108-89-2	N/R	N/R	N/R	N/S	N/R	N/S	1-3/16	N/A	1-1/2	N/A
108-83-2	108-89-2	N/R	N/R	N/R	N/S	N/R	N/S	1-3/16	N/A	1-1/2	N/A
108-83-2	108-89-2	108-83-2	108-89-2	N/R	134-103	134-102	N/S	1-5/16	1-3/8	1-11/16	1-11/16
108-83-2	108-89-2	N/R	N/R	N/R	N/S	N/R	N/S	1-3/16	N/A	1-1/2	N/A
108-83-2	108-89-2	108-90-2	108-90-2	108-27-2	134-101	N/A	N/A	1-1/4	1-5/16	1-11/16	1-11/16
108-83-2	108-89-2	108-90-2	108-90-2	108-27-2	134-101	N/A	N/S	1-3/8	1-7/16	1-11/16	1-11/16
108-83-2	108-89-2	108-90-2	108-90-2	108-27-2	134-101	N/A	N/S	1-3/8	1-7/16	1-11/16	1-11/16
108-83-2	108-89-2	108-83-2	108-89-2	N/R	N/S	N/S	N/S	1-1/4	1-5/16	1-9/16	1-9/16
108-83-2	108-89-2	108-83-2	108-89-2	N/R	134-103	N/S	N/S	1-9/16	1-9/16	1-3/4	1-3/4
108-83-2	108-89-2	108-83-2	108-89-2	N/R	N/S	N/S	N/S	1-3/8	1-7/16	1-11/16	1-11/16
N/R	N/R	108-90-2	108-90-2	108-13-2	N/S	N/R	N/S	1-9/16	N/A	1-3/4	N/A

Carb Number	Carb Model Number	CFM	Renew Kit	Trick Kit	Needle & Seat	Primary Main Jet	Secondary Main Jet or plate	Primary Metering Block	Secondary Metering Block	Power Valve	Pump Nozzle Size, Primary	Secondary Nozzle Size or Spring Color
R4412	2300	500	37-474	37-933	6-504	122-73	N/R	134-137	N/R	125-50	0.028	N/R
R4412-1	2300	500	37-474	37-933	6-504	122-73	N/R	134-137	N/R	125-50	0.028	N/R
R4412-2	2300	500	37-474	37-933	6-504	122-73	N/R	134-137	N/R	125-50	0.028	N/R
R4412-3	2300	500	37-474	37-933	6-504	122-73	N/R	134-137	N/R	125-50	0.028	N/R
R4412-4	2300	500	37-474	37-933	6-504	122-73	N/R	134-137	N/R	125-50	0.028	N/R
R4412-5	2300	500	37-474	37-933	6-504	122-73	N/R	134-137	N/R	125-50	0.028	N/R
R4452-1	4160	600	37-119	37-933	6-506	122-63	134-39	N/S	N/S	125-85	0.031	Purple
R4490	4150	780	37-1539	37-933	6-504	122-70	122-76	N/S	N/S	125-85 (12)	0.025	Yellow
R4514-1	4150	700	37-1537	37-933	6-504	122-66	122-79	N/S	N/S	125-65	0.029	Yellow
R4548	4160	450	37-119	37-933	6-506	122-57	N/S	N/S	N/S	N/S	0.031	Brown
R4555	4150	780	37-1539	37-933	6-504	122-70	122-76	N/S	N/S	125-85 (12)	0.025	Yellow
R4575	4500	1050	37-1539	37-933	6-504	122-84	122-84	N/S	N/S	125-65 (15)	0.035	0.035
R4609	4150	730	37-1537	37-933	6-504	122-66	122-79	N/S	N/S	125-65	0.029	Yellow
R4628	4150	780	37-1537	37-933	6-504	122-70	122-83	N/S	N/S	125-85	0.026	Yellow
R4647	4150	735	37-1537	37-933	6-504	122-64	122-82	N/S	N/S	125-85	0.031	Purple
R4653	4150	780	37-1537	37-933	6-504	122-71	122-82	N/S	N/S	125-65	0.026	Purple
R4670	2300	350	37-1537	37-933	6-504	122-62	N/R	N/S	N/R	125-65	0.031	N/R
R4672	2300	500	37-1537	37-933	6-504	N/R	N/S	N/R	N/R	N/R	N/R	Yellow
R4691-2	2110	300	N/A	N/A	N/S	122-63	N/R	N/R	N/R	N/R	0.021	N/R
R4742	4150	600	37-1539	37-933	6-504	122-63	122-72	N/S	N/S	N/S	0.031	Purple
R4773	4160	450	703-1	N/A	6-506	122-58	N/S	N/S	N/S	125-85	0.025	ORANGE
R4776	4150	600	37-485	37-933	6-504	122-69	122-71	N/S	N/A	125-65	0.025	0.032
R4776-1	4150	600	37-485	37-933	6-504	122-66	122-76	N/S	N/A	125-65	0.028	0.032
R4776-2	4150	600	37-485	37-933	6-504	122-66	122-76	N/A	N/A	125-65	0.028	0.032
R4776-3	4150	600	37-485	37-933	6-504	122-66	122-73	N/A	N/A	125-65	0.028	0.032
R4776-4	4150	600	37-485	37-933	6-504	122-66	122-73	N/A	N/A	125-65	0.028	0.032
R4776-5	4150	600	37-485	37-933	6-504	122-66	122-73	N/A	N/A	125-65	0.028	0.032
R4776-6	4150	600	37-485	37-933	6-504	122-66	122-73	N/A	N/A	125-65	0.028	0.032
R4776-7	4150	600	37-485	37-933	6-504	122-66	122-73	134-63	134-64	125-65	0.028	0.032
R4777	4150	650	37-485	37-933	6-504	122-71	122-76	N/S	N/A	125-65	0.025	0.025
R4777-1	4150	650	37-485	37-933	6-504	122-67	122-76	N/S	N/A	125-65	0.028	0.028
R4777-2	4150	650	37-485	37-933	6-504	122-67	122-76	134-150	N/A	125-65	0.028	0.028
R4777-3	4150	650	37-485	37-933	6-504	122-67	122-73	134-150	N/A	125-65	0.028	0.028
R4777-4	4150	650	37-485	37-933	6-504	122-67	122-73	134-150	N/A	125-65	0.028	0.028
R4777-5	4150	650	37-485	37-933	6-504	122-67	122-73	N/A	N/A	125-65	0.028	0.028
R4777-6	4150	650	37-485	37-933	6-504	122-67	122-73	N/A	N/A	125-65	0.028	0.028
R4777-7	4150	650	37-485	37-933	6-504	122-67	122-73	N/A	N/A	125-65	0.028	0.028
R4778	4150	700	37-485	37-933	6-504	122-66	122-71	N/S	N/S	125-65	0.025	0.032
R4778-1	4150	700	37-485	37-933	6-504	122-66	122-76	N/S	N/S	125-65	0.028	0.031
R4778-2	4150	700	37-485	37-933	6-504	122-66	122-76	N/S	N/S	125-65	0.028	0.031
R4778-3	4150	700	37-485	37-933	6-504	122-69	122-78	N/S	N/S	125-65	0.028	0.031
R4778-4	4150	700	37-485	37-933	6-504	122-69	122-78	N/S	N/S	125-65	0.028	0.031
R4778-5	4150	700	37-485	37-933	6-504	122-69	122-78	N/A	N/A	125-65	0.028	0.031
R4778-6	4150	700	37-485	37-933	6-504	122-69	122-78	N/A	N/A	125-65	0.028	0.031
R4778-7	4150	700	37-485	37-933	6-504	122-69	122-78	N/A	N/A	125-65	0.028	0.031
R4778-8	4150	700	37-485	37-933	6-504	122-69	122-78	N/A	N/A	125-65	0.028	0.031
R4779	4150	750	37-485	37-933	6-504	122-75	122-76	N/S	N/S	125-85	0.025	0.031
R4779-1	4150	750	37-485	37-933	6-504	122-70	122-80	N/S	N/S	125-85	0.028	0.031
R4779-2	4150	750	37-485	37-933	6-504	122-70	122-80	134-155	N/S	125-65	0.028	0.031
R4779-3	4150	750	37-485	37-933	6-504	122-70	122-73	N/A	N/S	125-65	0.028	0.031
R4779-4	4150	750	37-485	37-933	6-504	122-70	122-80	N/A	N/S	125-65	0.028	0.031
R4779-5	4150	750	37-485	37-933	6-504	122-70	122-80	N/A	N/S	125-65	0.028	0.031
R4779-6	4150	750	37-485	37-933	6-504	122-71	122-80	N/A	N/A	125-65	0.028	0.031

Primary Bowl Gasket	Primary Metering Block Gasket	Secondary Bowl Gasket	Secondary Metering Block Gasket	Secondary Metering Plate Gasket	Primary Fuel Bowl	Secondary Fuel Bowl	Throttle Body & Shaft Assembly	Venturii Diameter Primary	Venturii Diameter Secondary	Throttle Bore Diameter Primary	Throttle Bore Diameter Secondary
108-83-2	108-89-2	N/R	N/R	N/R	134-103	N/R	112-2	1-3/8	N/R	1-11/16	N/R
108-83-2	108-89-2	N/R	N/R	N/R	134-103	N/R	112-2	1-3/8	N/R	1-11/16	N/R
108-83-2	108-89-2	N/R	N/R	N/R	134-103	N/R	112-2	1-3/8	N/R	1-11/16	N/R
108-83-2	108-89-2	N/R	N/R	N/R	134-103S	N/R	112-2	1-3/8	N/R	1-11/16	N/R
108-83-2	108-89-2	N/R	N/R	N/R	134-103	N/R	112-2	1-3/8	N/R	1-11/16	N/R
108-83-2	108-89-2	N/R	N/R	N/R	134-103	N/R	112-2	1-3/8	N/R	1-11/16	N/R
108-83-2	108-89-2	108-90-2	108-90-2	108-27-2	N/A	134-105	N/S	1-1/4	1-5/16	1-9/16	1-9/16
108-83-2	108-89-2	108-83-2	108-89-2	N/R	134-103	N/S	N/S	1-3/8	1-7/16	1-11/16	1-11/16
108-83-2	108-89-2	108-83-2	108-89-2	N/R	N/S	N/S	N/S	1-3/8	1-7/16	1-11/16	1-11/16
108-83-2	108-89-2	108-90-2	108-90-2	108-27-2	N/A	134-105	N/S	1-3/32	1-3/32	1-1/2	1-1/2
108-83-2	108-89-2	108-83-2	108-89-2	N/R	134-103	N/S	N/S	1-3/8	1-7/16	1-11/16	1-11/16
108-83-2	108-89-2	108-83-2	108-89-2	N/R	134-108	134-112	N/S	1-11/16	1-11/16	2	2
108-83-2	108-89-2	108-83-2	108-89-2	N/R	N/S	N/S	N/S	1-3/8	1-7/16	1-11/16	1-11/16
108-83-2	108-89-2	108-83-2	108-89-2	N/R	N/S	N/S	N/S	1-3/8	1-7/16	1-11/16	1-11/16
108-83-2	108-89-2	108-83-2	108-89-2	N/R	N/S	N/S	N/S	1-3/8	1-7/16	1-11/16	1-11/16
108-83-2	108-89-2	108-83-2	108-89-2	N/R	N/S	N/S	N/S	1-3/8	1-7/16	1-11/16	1-11/16
108-83-2	108-89-2	N/R	N/R	N/R	N/S	N/R	N/S	1-3/16	N/A	1-1/2	N/A
N/R	N/R	108-90-2	108-90-2	108-13-2	N/R	N/S	N/S	1-9/16	N/A	1-3/4	N/A
N/R	N/R	N/R	N/R	N/R	N/R	N/R	N/R	1/5/32	N/R	1-7/16	N/R
108-83-2	108-89-2	108-83-2	108-89-2	N/R	N/S	N/S	N/S	1-1/4	1-5/16	1-9/16	1-9/16
N/A	N/A	N/A	N/A	108-27-2	N/S	N/S	N/S	1-3/32	1-3/32	1-1/2	1-1/2
108-83-2	108-89-2	108-83-2	108-89-2	N/R	134-103	134-104	N/S	1-1/4	1-5/16	1-9/16	1-9/16
108-83-2	108-89-2	108-83-2	108-89-2	N/R	134-103	134-104	N/S	1-1/4	1-5/16	1-9/16	1-9/16
108-83-2	108-89-2	108-83-2	108-89-2	N/R	134-103	134-104	N/A	1-1/4	1-5/16	1-9/16	1-9/16
108-83-2	108-89-2	108-83-2	108-89-2	N/R	134-103	134-104	N/A	1-1/4	1-5/16	1-9/16	1-9/16
108-83-2	108-89-2	108-83-2	108-89-2	N/R	134-103	134-104	N/A	1-1/4	1-5/16	1-9/16	1-9/16
108-83-2	108-89-2	108-83-2	108-89-2	N/R	134-103S	134-104S	N/A	1-1/4	1-5/16	1-9/16	1-9/16
108-83-2	108-89-2	108-83-2	108-89-2	N/R	134-103	134-104	N/A	1-1/4	1-5/16	1-9/16	1-9/16
108-83-2	108-89-2	108-83-2	108-89-2	N/R	134-103	134-104	112-118	1-1/4	1-5/16	1-9/16	1-9/16
108-83-2	108-89-2	108-83-2	108-89-2	N/R	134-103	134-104	N/S	1-1/4	1-5/16	1-11/16	1-11/16
108-83-2	108-89-2	108-83-2	108-89-2	N/R	134-103	134-104	N/S	1-1/4	1-5/16	1-11/16	1-11/16
108-83-2	108-89-2	108-83-2	108-89-2	N/R	134-103	134-104	112-17	1-1/4	1-5/16	1-11/16	1-11/16
108-83-2	108-89-2	108-83-2	108-89-2	N/R	134-103	134-104	112-17	1-1/4	1-5/16	1-11/16	1-11/16
108-83-2	108-89-2	108-83-2	108-89-2	N/R	134-103	134-104	112-17	1-1/4	1-5/16	1-11/16	1-11/16
108-83-2	108-89-2	108-83-2	108-89-2	N/R	134-103S	134-104S	N/A	1-1/4	1-5/16	1-11/16	1-11/16
108-83-2	108-89-2	108-83-2	108-89-2	N/R	134-103S	134-104S	N/A	1-1/4	1-5/16	1-11/16	1-11/16
108-83-2	108-89-2	108-83-2	108-89-2	N/R	134-103	134-104	N/A	1-1/4	1-5/16	1-11/16	1-11/16
108-83-2	108-89-2	108-83-2	108-89-2	N/R	134-103	134-104	N/S	1-5/16	1-3/8	1-11/16	1-11/16
108-83-2	108-89-2	108-83-2	108-89-2	N/R	134-103	134-104	N/S	1-5/16	1-3/8	1-11/16	1-11/16
108-83-2	108-89-2	108-83-2	108-89-2	N/R	134-103	134-104	N/A	1-5/16	1-3/8	1-11/16	1-11/16
108-83-2	108-89-2	108-83-2	108-89-2	N/R	134-103	134-104	N/A	1-5/16	1-3/8	1-11/16	1-11/16
108-83-2	108-89-2	108-83-2	108-89-2	N/R	134-103	134-104	N/A	1-5/16	1-3/8	1-11/16	1-11/16
108-83-2	108-89-2	108-83-2	108-89-2	N/R	134-103S	134-104S	N/A	1-5/16	1-3/8	1-11/16	1-11/16
108-83-2	108-89-2	108-83-2	108-89-2	N/R	134-103	134-104	N/A	1-5/16	1-3/8	1-11/16	1-11/16
108-83-2	108-89-2	108-83-2	108-89-2	N/R	134-103	134-104	N/A	1-5/16	1-3/8	1-11/16	1-11/16
108-83-2	108-89-2	108-83-2	108-89-2	N/R	134-103	134-104	N/A	1-5/16	1-3/8	1-11/16	1-11/16
108-83-2	108-89-2	108-83-2	108-89-2	N/R	134-103	134-104	N/S	1-3/8	1-3/8	1-11/16	1-11/16
108-83-2	108-89-2	108-83-2	108-89-2	N/R	134-103	134-104	N/S	1-3/8	1-3/8	1-11/16	1-11/16
108-83-2	108-89-2	108-83-2	108-89-2	N/R	134-103	134-104	N/S	1-3/8	1-3/8	1-11/16	1-11/16
108-83-2	108-89-2	108-83-2	108-89-2	N/R	134-103	134-104	N/S	1-3/8	1-3/8	1-11/16	1-11/16
108-83-2	108-89-2	108-83-2	108-89-2	N/R	134-103	134-104	N/S	1-3/8	1-3/8	1-11/16	1-11/16
108-83-2	108-89-2	108-83-2	108-89-2	N/R	134-103	134-104	N/A	1-3/8	1-3/8	1-11/16	1-11/16

Carb Number	Carb Model Number	CFM	Renew Kit	Trick Kit	Needle & Seat	Primary Main Jet	Secondary Main Jet or plate	Primary Metering Block	Secondary Metering Block	Power Valve	Pump Nozzle Size, Primary	Secondary Nozzle Size or Spring Color
R4779-7	4150	750	37-485	37-933	6-504	122-71	122-80	N/A	N/A	125-65	0.028	0.031
R4779-8	4150	750	37-485	37-933	6-504	122-71	122-80	N/A	N/A	125-65	0.028	0.031
R4779-9	4150	750	37-485	37-933	6-504	122-71	122-80	134-61	134-62	125-65	0.028	0.031
R4780	4150	800	37-485	37-933	6-504	122-72	122-76	N/S	N/S	(12,21)	0.031	0.031
R4780-1	4150	800	37-485	37-933	6-504	122-70	122-76	N/S	N/S	(12,21)	0.031	0.031
R4780-2	4150	800	37-485	37-933	6-504	122-70	122-85	N/S	N/S	125-65	0.031	0.031
R4780-3	4150	800	37-485	37-933	6-504	122-71	122-85	N/A	N/S	125-65	0.031	0.031
R4780-4	4150	800	37-485	37-933	6-504	122-71	122-85	N/A	N/S	125-65	0.031	0.031
R4780-5	4150	800	37-485	37-933	6-504	122-71	122-85	N/A	N/A	125-65	0.031	0.031
R4780-6	4150	800	37-485	37-933	6-504	122-71	122-85	N/A	N/A	125-65	0.031	0.031
R4780-7	4150	800	37-485	37-933	6-504	122-71	122-85	N/A	N/A	125-65	0.031	0.031
R4781	4150	850	37-485	37-933	6-504	122-80	122-80	N/S	N/S	125-65 (15)	0.035	0.025
R4781-1	4150	850	37-485	37-933	6-504	122-80	122-80	N/S	N/S	125-65 (15)	0.031	0.031
R4781-2	4150	850	37-485	37-933	6-504	122-80	122-80	N/A	N/S	125-65 (15)	0.031	0.031
R4781-3	4150	850	37-485	37-933	6-504	122-80	122-78	N/S	N/S	125-65 (15)	0.031	0.031
R4781-4	4150	850	37-485	37-933	6-504	122-80	122-78	N/S	N/S	125-65 (15)	0.031	0.031
R4781-5	4150	850	37-485	37-933	6-504	122-80	122-78	N/A	N/S	125-65 (15)	0.031	0.031
R4781-6	4150	850	37-485	37-933	6-504	122-80	122-78	N/A	N/A	125-65 (15)	0.031	0.031
R4781-7	4150	850	37-485	37-933	6-504	122-80	122-78	N/A	N/A	125-65 (15)	0.031	0.031
R4781-8	4150	850	37-485	37-933	6-504	122-80	122-78	134-65	134-66	125-65 (15)	0.031	0.031
R4782	2300	355	37-1537	37-933	6-504	122-64	N/R	N/S	N/R	125-65	0.031	N/R
R4783	2300	500	37-1537	37-933	6-504	122-82	N/R	N/R	N/S	N/R	0.028	N/R
R4788	4150	830	37-485	37-933	6-504	122-80	122-80	N/S	N/S	125-65 (B)	0.031	0.031
R4788-1	4150	830	37-485	37-933	6-504	122-80	122-80	N/S	N/S	125-65 (B)	0.031	0.031
R4790	2300	500	37-1537	37-933	6-504	N/R	N/S	N/R	N/R	N/R	N/R	Yellow
R4791	2300	350	37-1537	37-933	6-504	122-62	N/R	N/S	N/R	125-65	0.031	N/R
R4792	2300	350	37-1537	37-933	6-504	122-61	N/R	N/S	N/R	125-65	0.031	N/R
R4800-1	4150	780	37-1539	37-933	6-504	122-70	122-76	N/S	N/S	125-85 (12)	0.025	Yellow
R4801-1	4150	780	37-1539	37-933	6-504	122-70	122-76	N/S	N/S	128-85 (12)	0.025	Yellow
R4802-1	4150	780	37-1539	37-933	6-504	122-70	122-76	N/S	N/S	125-85 (12)	0.025	Yellow
R4803-1	4150	780	37-1539	37-933	6-504	122-70	122-76	N/S	N/S	125-85 (12)	0.025	Yellow
R6105	2300	500	3-396	N/A	6-504	N/R	N/S	N/S	N/S	N/R	N/R	N/S
R6106	2300	350	3-396	N/A	6-504	N/R	N/S	N/S	N/S	125-65	0.031	N/S
R6107	2300	500	3-396	N/A	6-504	N/R	N/S	N/S	N/S	N/R	N/R	N/S
R6109	4150	750	37-485	37-933	6-504	122-75	122-76	N/S	N/S	125-85	0.025	0.032
R6129	4150	780	37-1537	37-933	6-504	122-70	122-82	N/S	N/S	125-65	0.026	Purple
R6150	2300	300	3-888	N/A	6-506	122-59	N/S	N/S	N/S	125-25	0.028	N/S
R6151	4160	600	703-1	N/A	6-506	122-66	34R9716-3	N/S	N/S	125-105	0.025	PURPLE
R6152	4160	600	703-1	N/A	6-506	122-66	N/S	N/S	N/S	125-85	0.025	PURPLE
R6210-1	4165	650	37-605	37-933	(16,17)	N/A	N/A	N/S	N/S	(14,15)	0.025	0.037
R6210-2	4165	650	37-605	37-933	(16,17)	N/A	122-83	N/S	N/S	125-85	0.025	0.037
R6210-3	4165	650	37-605	37-933	(16,17)	N/A	122-83	N/S	N/S	125-85	0.025	0.037
R6211	4165	800	37-605	37-933	16,17)	122-62	122-85	N/S	N/S	(14,15)	0.025	0.037
R6211-1	4165	800	37-605	37-933	(16,17)	N/A	122-85	N/S	N/S	(14,15)	0.025	0.037
R6212	4165	800	37-1537	37-933	6-504	122-63	122-86	N/S	N/S	(14,15)	0.025	0.037
R6213	4165	800	37-1537	37-933	6-504	122-62	122-85	N/S	N/S	(14,15)	0.025	0.037
R6214	4500	1150	N/A	N/A	6-504	122-95	122-95	N/S	N/S	N/R	0.026	0.026
R6238-1	4150	780	37-1539	37-933	6-504	122-68	122-73	N/S	N/S	125-65 (12)	0.025	Yellow
R6239-1	4150	780	37-1539	37-933	6-504	122-68	122-73	N/S	N/S	125-65 (12)	0.025	Yellow
R6244-1	2110	200	N/A	N/A	N/A	122-47	N/R	N/R	N/R	N/R	0.021	N/R
R6262	4165	800	37-605	37-933	(16,17)	122-62	122-85	N/S	N/S	(14,15)	0.025	0.037
R6270-1	4160	600	37-1536	37-933	N/S	122-64	N/S	N/S	N/S	125-85	0.032	Orange
R6291	4160	600	37-119	37-933	6-506	122-62	134-39	N/S	N/S	125-85	0.031	Purple

Primary Bowl Gasket	Primary Metering Block Gasket	Secondary Bowl Gasket	Secondary Metering Block Gasket	Secondary Metering Plate Gasket	Primary Fuel Bowl	Secondary Fuel Bowl	Throttle Body & Shaft Assembly	Venturii Diameter Primary	Venturii Diameter Secondary	Throttle Bore Diameter Primary	Throttle Bore Diameter Secondary
108-83-2	108-89-2	108-83-2	108-89-2	N/R	134-103S	134-104S	N/A	1-3/8	1-3/8	1-11/16	1-11/16
108-83-2	108-89-2	108-83-2	108-89-2	N/R	134-103S	134-104S	N/A	1-3/8	1-3/8	1-11/16	1-11/16
108-83-2	108-89-2	108-83-2	108-89-2	N/R	134-103	134-104	112-116	1-3/8	1-3/8	1-11/16	1-11/16
108-83-2	108-89-2	108-83-2	108-89-2	N/R	134-103	134-104	N/S	1-3/8	1-7/16	1-11/16	1-11/16
108-83-2	108-89-2	108-83-2	108-89-2	N/R	134-103	134-104	N/S	1-3/8	1-7/16	1-11/16	1-11/16
108-83-2	108-89-2	108-83-2	108-89-2	N/R	134-103	134-104	N/A	1-3/8	1-7/16	1-11/16	1-11/16
108-83-2	108-89-2	108-83-2	108-89-2	N/R	134-103	134-104	N/A	1-3/8	1-7/16	1-11/16	1-11/16
108-83-2	108-89-2	108-83-2	108-89-2	N/R	134-103	134-104	N/A	1-3/8	1-7/16	1-11/16	1-11/16
108-83-2	108-89-2	108-83-2	108-89-2	N/R	134-103S	134-104S	N/A	1-3/8	1-7/16	1-11/16	1-11/16
108-83-2	108-89-2	108-83-2	108-89-2	N/R	134-103	134-104	N/A	1-3/8	1-7/16	1-11/16	1-11/16
108-83-2	108-89-2	108-83-2	108-89-2	N/R	134-103	134-104	N/A	1-3/8	1-7/16	1-11/16	1-11/16
108-83-2	108-89-2	108-83-2	108-89-2	N/R	134-103	134-104	N/S	1-9/16	1-9/16	1-3/4	1-3/4
108-83-2	108-89-2	108-83-2	108-89-2	N/R	134-103	134-104	N/S	1-9/16	1-9/16	1-3/4	1-3/4
108-83-2	108-89-2	108-83-2	108-89-2	N/R	134-103	134-104	N/S	1-9/16	1-9/16	1-3/4	1-3/4
108-83-2	108-89-2	108-83-2	108-89-2	N/R	134-103	134-104	N/S	1-9/16	1-9/16	1-3/4	1-3/4
108-83-2	108-89-2	108-83-2	108-89-2	N/R	134-103	134-104	N/S	1-9/16	1-9/16	1-3/4	1-3/4
108-83-2	108-89-2	108-83-2	108-89-2	N/R	134-103	134-104	N/A	1-9/16	1-9/16	1-3/4	1-3/4
108-83-2	108-89-2	108-83-2	108-89-2	N/R	134-103S	134-104S	N/A	1-9/16	1-9/16	1-3/4	1-3/4
108-83-2	108-89-2	108-83-2	108-89-2	N/R	134-103	134-104	N/A	1-9/16	1-9/16	1-3/4	1-3/4
108-83-2	108-89-2	108-83-2	108-89-2	N/R	134-103	134-104	112-119	1-9/16	1-9/16	1-3/4	1-3/4
108-83-2	108-89-2	N/R	N/R	N/R	N/S	N/R	N/S	1-3/16	N/R	1-1/2	N/R
108-83-2	108-89-2	N/R	N/R	N/R	N/S	N/R	N/S	1-9/16	N/R	1-3/4	N/R
108-83-2	108-89-2	108-83-2	108-89-2	N/R	134-103	134-104	N/S	1-9/16	1-9/16	1-11/16	1-11/16
108-83-2	108-89-2	108-83-2	108-89-2	N/R	134-103	134-104	N/S	1-9/16	1-9/16	1-11/16	1-11/16
N/R	N/R	108-90-2	108-90-2	108-13-2	N/R	N/S	N/S	1-9/16	N/R	1-3/4	N/R
108-83-2	108-89-2	N/R	N/R	N/R	N/S	N/R	N/S	1-3/16	N/R	1-1/2	N/R
108-83-2	108-89-2	N/R	N/R	N/R	N/S	N/R	N/S	1-3/16	N/R	1-1/2	N/R
108-83-2	108-89-2	108-83-2	108-89-2	N/R	134-103	N/S	N/S	1-3/8	1-7/16	1-11/16	1-11/16
108-83-2	108-89-2	108-83-2	108-89-2	N/R	134-103	N/S	N/S	1-3/8	1-7/16	1-11/16	1-11/16
108-83-2	108-89-2	108-83-2	108-89-2	N/R	134-103	N/S	N/S	1-3/8	1-7/16	1-11/16	1-11/16
108-83-2	108-89-2	108-83-2	108-89-2	N/R	134-103	N/S	N/S	1-3/8	1-7/16	1-11/16	1-11/16
N/A	N/A	N/A	N/A	108-13-2	N/S	N/S	N/S	1-9/16	N/R	1-3/4	N/R
N/A	N/A	N/A	N/A	N/A	N/S	N/S	N/S	1-3/16	N/R	1-1/2	N/R
N/A	N/A	N/A	N/A	108-13-2	N/S	N/S	N/S	1-9/16	N/R	1-3/4	N/R
108-83-2	108-89-2	108-83-2	108-89-2	N/R	N/S	N/S	N/S	1-3/8	1-3/8	1-11/16	1-11/16
108-83-2	108-89-2	108-83-2	108-89-2	N/R	N/S	N/S	N/S	1-3/8	1-7/16	1-11/16	1-11/16
N/A	N/A	N/A	N/A	N/A	N/S	N/S	N/S	1-3/16	N/R	1-1/2	N/R
N/A	N/A	N/A	N/A	108-27-2	N/S	N/S	N/S	1-1/4	1-5/16	1-9/16	1-9/16
N/A	N/A	N/A	N/A	108-27-2	N/S	N/S	N/S	1-1/4	1-5/16	1-9/16	1-9/16
108-92-2	108-91-2	108-92-2	108-91-2	N/R	134-110	N/A	N/S	1-5/32	1-3/8	1-3/8	2
108-92-2	108-91-2	108-92-2	108-91-2	N/R	134-110	N/A	N/S	1-5/32	1-3/8	1-3/8	2
108-92-2	108-91-2	108-92-2	108-91-2	N/R	134-110	N/A	N/A	1-5/32	1-3/8	1-3/8	2
108-92-2	108-91-2	108-92-2	108-91-2	N/R	134-110	N/A	N/S	1-5/32	1-23/32	1-3/8	2
108-92-2	108-91-2	108-92-2	108-91-2	N/R	134-110	N/A	N/S	1-5/32	1-23/32	1-3/8	2
108-92-2	108-91-2	108-92-2	108-91-2	N/R	N/S	N/S	N/S	1-5/32	1-23/32	1-3/8	2
108-92-2	108-91-2	108-92-2	108-91-2	N/R	N/S	N/S	N/S	1-5/32	1-23/32	1-3/8	2
108-83-2	108-36-2	108-83-2	108-36-2	N/R	134-108	134-112	N/S	1-13/16	1-13/16	2	2
108-83-2	108-89-2	108-83-2	108-89-2	N/R	N/S	N/S	N/S	1-3/8	1-7/16	1-11/16	1-11/16
108-83-2	108-89-2	108-83-2	108-89-2	N/R	N/S	N/S	N/S	1-3/8	1-7/16	1-11/16	1-11/16
N/R	N/R	N/R	N/R	N/R	N/R	N/R	N/R	1-5/16	N/R	1-7/16	N/R
108-92-2	108-91-2	108-92-2	108-91-2	N/R	134-110	N/A	N/S	1-13/16	1-23/32	1-3/8	2
108-83-2	108-34-2	108-90-2	108-90-2	108-27-2	N/S	N/S	N/S	1-1/4	1-5/16	1-9/16	1-9/16
108-83-2	108-91-2	108-90-2	108-90-2	108-27-2	N/S	N/S	N/S	1-1/4	1-5/16	1-5/16	1-9/16

Carb Number	Carb Model Number	CFM	Renew Kit	Trick Kit	Needle & Seat	Primary Main Jet	Secondary Main Jet or plate	Primary Metering Block	Secondary Metering Block	Power Valve	Pump Nozzle Size, Primary	Secondary Nozzle Size or Spring Color
R6299-1	4160	390	37-1539	37-933	6-506	122-50	N/A	N/S	N/S	N/A	0.025	Plain
R6319	2300	300	3-888	N/A	6-506	122-60	N/S	N/S	N/S	125-50	0.028	N/S
R6361	4150	650	3-1184	N/A	6-504	122-72	122-84	N/S	N/S	125-85	0.026	YELLOW
R6407	4160	450	703-1	N/A	6-506	122-58	N/S	N/S	N/S	125-85	0.021	PLAIN
R6425	2300	650	N/A	N/A	6-504	122-82	N/R	N/S	N/S	125-65	0.031	N/R
R6464	4500	1050	37-1539	37-933	6-504	122-88	122-88	N/S	N/R	N/R	0.035	0.035
R6468-1	4165	650	37-605	37-933	(16,17)	122-60	122-83	N/S	N/S	125-85	0.025	0.037
R6468-2	4165	650	37-605	37-933	(16,17)	N/A	122-83	N/S	N/A	125-85	0.025	0.037
R6497	4165	650	37-605	37-933	(16,17)	N/A	N/A	N/S	N/S	(14,15)	0.025	0.037
R6498	4165	650	37-605	37-933	(16,17)	N/A	N/A	N/S	N/S	(14,15)	0.025	0.037
R6499	4165	650	37-1537	N/A	6-504	122-60	122-63	N/S	N/S	(14,15)	0.025	0.037
R6512	4165	650	37-605	37-933	(16,17)	122-60	122-60	N/S	N/S	(14,15)	0.025	0.037
R6520	4160	600	37-119	37-933	6-506	122-62	134-39	N/S	N/S	125-85	0.031	Purple
R6528	4165	650	37-605	37-933	(16,17)	122-61	122-60	N/S	N/S	(14,15)	0.025	0.037
R6619-1	4160	600	37-720	37-933	6-506	N/A	134-39	N/S	134-39	125-65	0.031	Black
R6647	4150	750	N/A	N/A	6-504	122-68	122-70	N/S	N/S	125-85 (12)	0.025	Yellow
R6708	4150	650	37-1539	37-933	6-504	N/A	N/A	N/S	134-39	(21,22)	0.025	0.037
R6708-1	4150	650	37-1539	37-933	6-504	N/A	122-85	N/S	N/S	125-65	0.025	0.037
R6709	4150	750	37-1539	37-933	6-504	N/A	122-76	N/S	N/S	(21,22)	0.025	0.037
R6710	4165	800	37-1537	37-933	6-504	122-63	122-86	N/S	N/S	(21,22)	0.025	0.037
R6711	4165	650	37-605	37-933	(16,17)	N/A	N/A	N/S	N/S	(21,22)	0.025	0.028
R6772	4165	650	37-605	37-933	(16,17)	N/A	N/A	N/S	N/S	(14,15)	0.025	0.04
R6773	4165	650	37-605	37-933	(16,17)	N/A	N/A	N/S	N/S	(14,15)	0.025	0.04
R6774	4165	650	37-605	37-933	(16,17)	N/A	N/A	N/S	N/S	(14,15)	0.025	0.037
R6846	2300	300	N/A	N/A	6-511	122-60	N/S	N/S	N/S	125-50	0.028	N/S
R6853	4165	650	37-605	37-933	(16,17)	122-60	122-62	N/S	N/S	(14,15)	0.025	0.037
R6895	4150	390	37-1536	37-933	6-504	122-50	122-62	N/S	N/S	125-85	0.025	0.025
R6909	4160	600	37-119	37-933	6-506	N/A	134-39	N/S	134-39	125-65	0.031	Black
R6910	4165	800	37-1537	37-933	6-504	N/A	122-86	N/S	N/S	(14,15)	0.025	0.037
R6919	4160	600	37-1536	37-933	6-506	N/A	134-39	N/S	134-39	125-206	0.031	Black
R6946-1	4160	600	3-1012	N/A	6-504	N/A	N/S	N/S	N/S	N/A	0.025	Plain
R6947	4160	600	3-1012	N/A	6-504	N/A	N/S	N/S	N/S	125-206	0.025	Plain
R6979	4160	600	N/A	37-933	6-506	N/A	134-39	N/S	134-39	125-85	0.031	Black
R6979-1	4160	600	N/A	37-933	6-506	N/A	134-39	N/S	134-39	125-208	0.031	Black
R6989	4160	600	37-1536	37-933	6-506	N/A	134-39	N/S	134-39	125-206	0.031	Black
R7001	4165	650	N/A	37-933	(16,17)	N/A	N/A	N/S	N/S	(15,24)	0.025	0.037
R7002-1	4175	650	37-1537	37-933	(16,17)	N/A	134-21	N/S	134-21	125-85	0.025	Black
R7004-1	4175	650	37-1537	37-933	(16,17)	N/A	N/A	N/S	N/A	N/A	0.025	Plain
R7004-2	4175	650	37-1537	37-933	(16,17)	N/A	N/S	N/S	N/S	N/A	0.025	Plain
R7005-1	4175	650	37-1537	37-933	(16,17)	N/A	N/A	N/S	N/A	N/A	0.025	Plain
R7005-2	4175	650	37-1537	37-933	(16,17)	N/A	N/S	N/S	N/S	N/A	0.025	Plain
R7006-1	4175	650	37-1537	37-933	(16,17)	N/A	N/A	N/S	N/A	N/A	0.025	Plain
R7006-2	4175	650	37-1537	37-933	(16,17)	N/A	N/S	N/S	N/S	N/A	0.025	Plain
R7009-1	4160	600	37-1536	37-933	6-506	N/A	134-39	N/S	134-39	125-206	0.031	Black
R7010	4160	780	37-1537	37-933	6-506	N/A	N/S	N/S	N/S	125-65	0.025	Black
R7036	2300	300	703-32	N/A	6-511	122-60	N/S	N/S	N/S	125-50	0.028	N/S
R7053-1	4160	600	37-119	37-933	6-506	N/A	134-39	N/S	N/S	125-85	0.031	Purple
R7054	4165	650	37-605	37-933	(16,17)	N/A	N/A	N/S	N/S	(14,15)	0.025	0.037
R7128	4160	650	703-33	N/A	6-511	122-73	N/S	N/S	N/S	125-65	0.026	YELLOW
R7154	4160	600	37-119	37-933	6-506	122-62	N/S	N/S	N/S	125-85	0.031	Purple
R7159	4160	450	703-33	N/A	6-511	122-59	134-8	N/S	N/S	125-85	0.021	PLAIN
R7163	4160	600	703-33	N/A	6-511	122-66	N/S	N/S	N/S	125-25	0.025	PURPLE
R7320	4500	1150	37-1539	37-933	6-518-2	122-96	122-96	N/S	N/S	122-65	0.035	0.035

Primary Bowl Gasket	Primary Metering Block Gasket	Secondary Bowl Gasket	Secondary Metering Block Gasket	Secondary Metering Plate Gasket	Primary Fuel Bowl	Secondary Fuel Bowl	Throttle Body & Shaft Assembly	Venturii Diameter Primary	Venturii Diameter Secondary	Throttle Bore Diameter Primary	Throttle Bore Diameter Secondary
108-83-2	108-89-2	108-90-2	108-90-2	108-28-2	N/S	N/S	N/S	1-1/16	1-1/16	1-7/16	1-7/16
N/A	N/A	N/A	N/A	N/A	N/S	N/S	N/S	1-3/16	N/R	1-1/2	N/R
N/A	N/A	N/A	N/A	N/A	134-108	N/S	N/S	1-3/8	1-7/16	1-11/16	1-11/16
N/A	N/A	N/A	N/A	108-27-2	N/S	134-105	N/S	1-3/32	1-3/32	1-1/2	1-1/2
108-92-2	108-35-2	N/R	N/R	N/R	N/S	N/S	N/S	1-7/16	N/R	1-3/4	N/R
108-83-2	108-36-2	108-83-2	108-36-2	N/R	134-108	134-112	N/S	1-11/16	1-11/16	2	2
108-92-2	108-91-2	108-92-2	108-91-2	N/R	134-110	N/A	N/S	1-5/32	1-3/8	1-3/8	2
108-92-2	108-91-2	108-92-2	108-91-2	N/R	134-110	N/A	N/S	1-5/32	1-3/8	1-3/8	2
108-92-2	108-91-2	108-92-2	108-91-2	N/R	134-110	N/A	N/S	1-5/32	1-3/8	1-3/8	2
108-92-2	108-91-2	108-92-2	108-91-2	N/R	134-110	N/A	N/S	1-5/32	1-3/8	1-3/8	2
108-92-2	108-91-2	108-92-2	108-91-2	N/R	N/S	N/S	N/S	1-5/32	1-3/8	1-3/8	2
108-92-2	108-91-2	108-92-2	108-91-2	N/R	134-110	N/A	N/S	1-5/32	1-3/8	1-3/8	2
108-83-2	108-91-2	108-90-2	108-90-2	108-27-2	N/S	N/S	N/S	1-1/4	1-5/16	1-9/16	1-9/16
108-92-2	108-91-2	108-92-2	108-91-2	N/R	134-110	N/A	N/S	1-5/32	1-3/8	1-3/8	2
108-83-2	108-91-2	108-90-2	108-90-2	108-27-2	N/A	134-105	N/S	1-1/4	1-5/16	1-9/16	1-9/16
108-83-2	108-89-2	108-83-2	108-89-2	N/R	N/S	N/S	N/S	1-3/8	1-7/16	1-11/16	1-11/16
108-92-2	108-91-2	108-92-2	108-91-2	N/R	N/S	N/S	N/S	1-3/32	1-9/16	1-1/2	1-3/4
108-92-2	108-91-2	108-92-2	108-91-2	N/R	N/S	N/S	N/S	1-3/32	1-9/16	1-1/2	1-3/4
108-92-2	108-91-2	108-92-2	108-91-2	N/R	N/S	N/S	N/S	1-1/4	1-9/16	1-1/2	1-3/4
108-92-2	108-91-2	108-92-2	108-91-2	N/R	N/S	N/S	N/S	1-5/32	1-23/32	1-3/8	2
108-92-2	108-91-2	108-92-2	108-91-2	N/R	N/S	N/A	N/S	1-5/32	1-3/8	1-3/8	2
108-92-2	108-91-2	108-92-2	108-91-2	N/R	134-110	N/A	N/S	1-5/32	1-3/8	1-3/8	2
108-92-2	108-91-2	108-92-2	108-91-2	N/R	134-110	N/A	N/S	1-5/32	1-3/8	1-3/8	2
108-92-2	108-91-2	108-92-2	108-91-2	N/R	134-110	N/A	N/S	1-5/32	1-3/8	1-3/8	2
N/A	N/A	N/A	N/A	N/A	N/S	N/S	N/S	1-3/16	N/R	1-1/2	N/R
108-92-2	108-91-2	108-92-2	108-91-2	N/R	134-110	N/A	N/S	1-5/32	1-3/8	1-3/8	2
108-83-2	108-89-2	108-83-2	108-89-2	N/R	134-103	134-104	N/S	1-1/16	1-1/16	1-7/16	1-7/16
108-83-2	108-91-2	108-90-2	108-90-2	108-27-2	N/S	134-105	N/S	1-1/4	1-5/16	1-9/16	1-9/16
108-92-2	108-91-2	108-92-2	108-91-2	N/R	N/S	N/S	N/S	1-5/32	1-23/32	1-3/8	2
108-83-2	108-91-2	108-90-2	108-90-2	108-27-2	N/A	134-105	N/S	1-1/4	1-5/16	1-9/16	1-9/16
108-83-2	108-91-2	108-90-2	108-90-2	108-27-2	N/S	N/S	N/S	1-3/16	1-1/4	1-1/2	1-1/2
108-83-2	108-91-2	108-90-2	108-90-2	108-27-2	N/S	N/S	N/S	1-3/16	1-1/4	1-1/2	1-1/2
108-83-2	108-91-2	108-90-2	108-90-2	108-27-2	134-101	134-105	N/S	1-1/4	1-5/16	1-9/16	1-9/16
108-83-2	108-91-2	108-90-2	108-90-2	108-27-2	134-101	134-105	N/S	1-1/4	1-5/16	1-9/16	1-9/16
108-83-2	108-91-2	108-90-2	108-90-2	108-27-2	N/A	134-105	N/S	1-1/4	1-5/16	1-9/16	1-9/16
108-92-2	108-91-2	108-92-2	108-91-2	N/R	134-110	N/A	N/S	1-5/32	1-3/8	1-3/8	2
108-92-2	108-91-2	108-90-2	108-90-2	108-27-2	134-110	N/A	N/S	1-5/32	1-3/8	1-3/8	2
108-92-2	108-91-2	108-90-2	108-90-2	108-27-2	N/S	N/A	N/S	1-5/32	1-3/8	1-3/8	2
108-92-2	108-91-2	108-90-2	108-90-2	108-27-2	N/S	N/A	N/S	1-5/32	1-3/8	1-3/8	2
108-92-2	108-91-2	108-90-2	108-90-2	108-27-2	N/S	N/A	N/S	1-5/32	1-3/8	1-3/8	2
108-92-2	108-91-2	108-90-2	108-90-2	108-27-2	N/S	N/A	N/S	1-5/32	1-3/8	1-3/8	2
108-92-2	108-91-2	108-90-2	108-90-2	108-27-2	N/S	N/A	N/S	1-5/32	1-3/8	1-3/8	2
108-83-2	108-91-2	108-90-2	108-90-2	108-27-2	N/S	134-105	N/S	1-1/4	1-5/16	1-9/16	1-9/16
108-92-2	108-91-2	108-90-2	108-90-2	108-27-2	N/S	134-102	N/S	1-1/4	1-9/16	1-1/2	1-3/4
N/A	N/A	N/A	N/A	N/A	N/S	N/S	N/S	1-3/16	N/R	1-1/2	N/R
108-83-2	108-91-2	108-90-2	108-90-2	108-27-2	N/S	N/S	N/S	1-1/4	1-5/16	1-9/16	1-9/16
108-92-2	108-91-2	108-83-2	108-91-2	N/R	134-110	N/A	N/S	1-5/32	1-3/8	1-3/8	2
N/A	N/A	N/A	N/A	108-27-2	N/S	N/S	N/S	1-1/4	1-5/16	1-9/16	1-9/16
108-83-2	108-91-2	108-90-2	108-90-2	108-27-2	N/S	N/S	N/S	1-1/4	1-5/16	1-9/16	1-9/16
N/A	N/A	N/A	N/A	108-27-2	N/S	N/S	N/S	1-3/32	1-3/32	1-1/2	1-1/2
N/A	N/A	N/A	N/A	108-27-2	N/S	N/S	N/S	1-1/4	1-5/16	1-9/16	1-9/16
108-83-2	108-89-2	108-83-2	108-89-2	N/R	N/S	N/S	N/S	1.83	1.83	2	2

Carb Number	Carb Model Number	CFM	Renew Kit	Trick Kit	Needle & Seat	Primary Main Jet	Secondary Main Jet or plate	Primary Metering Block	Secondary Metering Block	Power Valve	Pump Nozzle Size, Primary	Secondary Nozzle Size or Spring Color
R7320-1	4500	1150	37-1539	37-933	6-518-2	122-95	122-95	N/S	N/S	N/R	0.031	0.035
R7320-2	4500	1150	37-1539	37-933	6-518-2	122-95	122-95	N/S	N/S	N/R	0.031	0.035
R7343	5200	230	N/A	N/A	N/S	N/S	N/S	N/S	N/S	N/S	0.02	N/R
R7344	5210	255	N/A	N/A	N/S	N/S	N/S	N/S	N/S	N/S	0.021	N/R
R7351	4175	650	37-1537	37-933	(16,17)	N/A	N/S	N/S	134-21	125-206	0.037	Black
R7397	4175	650	37-1537	37-933	(16,17)	N/A	134-21	N/S	134-21	125-206	0.037	Black
R7410	4150	340	37-1536	37-933	6-504	122-50	122-62	N/S	N/S	125-85	0.025	0.025
R7411	4150	370	37-1536	37-933	6-504	122-50	122-62	N/S	N/S	125-85	0.025	0.025
R7413	4160	600	37-119	37-933	6-506	N/A	134-39	N/S	N/S	125-85	0.031	Purple
R7448	2300	350	37-1536	37-933	6-504	122-61	N/A	134-203	N/R	125-85	0.031	N/R
R7448-1	2300	350	37-1536	37-933	6-504	122-61	N/A	134-203	N/R	125-85	0.031	N/R
R7454	4360	450	37-1540	N/A	N/S	N/A	N/A	N/R	N/R	N/S	0.028	N/R
R7455	4360	450	37-1540	N/A	N/S	N/A	N/A	N/R	N/R	N/S	0.028	N/R
R7456	4360	450	37-1540	N/A	N/S	N/A	N/A	N/R	N/R	N/S	0.028	N/R
R7555	4360	450	37-1540	N/A	N/S	N/A	N/A	N/R	N/R	N/S	0.028	N/R
R7556	4360	450	37-1540	N/A	N/S	N/A	N/A	N/R	N/R	N/S	0.028	N/R
R76650AABK	AL 4150	650	37-485	37-933	6-504	122-63	122-73	N/S	N/S	125-65	0.028	0.031
R76650AABL	AL 4150	650	37-485	37-933	6-504	122-63	122-73	N/S	N/S	125-65	0.028	0.031
R76650AARD	AL 4150	650	37-485	37-933	6-504	122-63	122-73	N/S	N/S	125-65	0.028	0.031
R76750AABK	AL 4150	750	37-485	37-933	6-504	122-72	122-80	N/S	N/S	125-65	0.028	0.031
R76750AABL	AL 4150	750	37-485	37-933	6-504	122-72	122-80	N/S	N/S	125-65	0.028	0.031
R76750AARD	AL 4150	750	37-485	37-933	6-504	122-72	122-80	N/S	N/S	125-65	0.028	0.031
R7850	4160	600	N/A	N/A	6-506	N/A	134-39	N/S	N/S	125-85	0.031	Plain
R7855	4175	650	37-1537	37-933	(16,17)	N/A	N/A	N/S	N/A	N/A	0.028	Plain
R7955	4360	450	37-1540	N/A	N/S	N/A	N/A	N/R	N/R	N/S	0.028	N/R
R7956	4360	450	37-1540	N/A	N/S	N/A	N/A	N/R	N/R	N/S	0.028	N/R
R7957	4360	450	37-1540	N/A	N/S	N/A	N/A	N/R	N/R	N/S	0.028	N/R
R7958	4360	450	37-1540	N/A	N/S	N/A	N/A	N/R	N/R	N/S	0.028	N/R
R7985	4160	600	37-1536	37-933	6-506	N/A	134-39	N/S	134-39	125-208	0.031	Black
R7986	4160	600	37-1536	37-933	6-506	N/A	134-39	N/S	134-39	125-208	0.031	Black
R7987	4160	600	37-1536	37-933	6-506	N/A	134-39	N/S	134-39	125-208	0.031	Black
R8001	4360	450	37-1540	N/A	N/S	N/A	N/A	N/R	N/R	N/S	0.028	N/R
R8002	4360	450	37-1540	N/A	N/S	N/A	N/A	N/R	N/R	N/S	0.028	N/R
R8003	4360	450	37-1540	N/A	N/S	N/A	N/A	N/R	N/R	N/S	0.028	N/R
R8004	4160	600	37-1536	37-933	6-506	N/A	134-39	N/S	134-39	125-208	0.031	Black
R8005	4160	600	37-1536	37-933	6-506	N/A	134-39	N/S	134-39	125-208	0.031	Black
R8006	4160	600	37-1536	37-933	6-506	N/A	134-39	N/S	134-39	125-208	0.031	Black
R8007	4160	390	37-720	37-933	6-506	122-51	N/A	N/A	N/A	125-65	0.025	Plain
R8059	4175	650	37-1537	37-933	(16,17)	N/A	134-21	N/S	134-21	125-206	0.037	Black
R8059-1	4175	650	37-1537	37-933	(16,17)	N/A	N/S	N/S	N/S	N/A	0.025	Black
R8060	4175	650	37-1537	37-933	(16,17)	N/A	134-21	N/S	134-21	125-206	0.037	Black
R8060-1	4175	650	37-1537	37-933	(16,17)	N/A	N/S	N/S	N/S	N/S	0.025	Black
R8082	4500	1050	37-1539	37-933	6-518-2	122-92	122-92	N/S	N/S	125-65	0.035	0.035
R8082-1	4500	1050	37-1539	37-933	6-504	122-88	122-88	N/S	N/S	125-65 (15)	0.035	0.035
R8082-2	4500	1050	37-1539	37-933	6-518-2	122-84	122-84	N/A	N/A	125-65 (15)	0.035	0.035
R8082-3	4500	1050	37-1539	37-933	6-518-2	122-84	122-84	N/A	N/A	125-65 (15)	0.035	0.035
R8123	4160	600	N/A	N/A	6-511	122-66	N/S	N/S	N/S	125-50	0.025	PURPLE
R8149	4360	450	37-1540	N/A	N/S	N/A	N/A	N/R	N/R	N/S	0.028	N/R
R8149-1	4360	450	37-1540	N/A	N/S	N/A	N/A	N/R	N/R	N/S	0.028	N/R
R8156	4150	750	37-485	37-933	6-504	122-70	122-83	134-155	N/S	125-65	0.028	0.031

Primary Bowl Gasket	Primary Metering Block Gasket	Secondary Bowl Gasket	Secondary Metering Block Gasket	Secondary Metering Plate Gasket	Primary Fuel Bowl	Secondary Fuel Bowl	Throttle Body & Shaft Assembly	Venturii Diameter Primary	Venturii Diameter Secondary	Throttle Bore Diameter Primary	Throttle Bore Diameter Secondary
108-83-2	108-89-2	108-83-2	108-89-2	N/R	134-108	134-112	N/S	1-13/16	1-13/16	2	2
108-83-2	108-89-2	108-83-2	108-89-2	N/R	134-108	134-112	N/S	1-13/16	1-13/16	2	2
N/R	N/R	N/R	N/R	N/R	N/R	N/R	N/R	1-1/25	1-1/16	1-7/25	1-7/16
N/R	N/R	N/R	N/R	N/R	N/R	N/R	N/R	1-1/33	1-1/16	1-1/4	1-7/25
108-92-2	108-91-2	108-90-2	108-90-2	108-27-2	134-110	N/A	N/S	1-13/64	1-13/32	1-3/8	2
108-92-2	108-91-2	108-90-2	108-90-2	108-27-2	134-110	N/A	N/S	1-13/64	1-13/32	1-3/8	2
108-83-2	108-89-2	108-83-2	108-89-2	N/R	N/S	N/S	N/S	1-1/16	1-1/16	1-7/16	1-7/16
108-83-2	108-89-2	108-83-2	108-89-2	N/R	N/S	N/S	N/S	1-1/16	1-1/16	1-7/16	1-7/16
108-83-2	108-91-2	108-90-2	108-90-2	108-27-2	N/S	N/S	N/S	1-1/4	1-5/16	1-9/16	1-9/16
108-83-2	108-89-2	N/R	N/R	N/R	134-103	N/R	N/A	1-3/16	N/R	1-1/2	N/R
108-83-2	108-89-2	N/R	N/R	N/R	134-103	N/R	N/A	1-3/16	N/R	1-1/2	N/R
N/S	N/R	N/R	N/R	N/R	N/R	N/R	N/S	1-1/16	1-3/16	1-3/8	1-7/16
N/S	N/R	N/R	N/R	N/R	N/R	N/R	N/S	1-1/16	1-3/16	1-3/8	1-7/16
N/S	N/R	N/R	N/R	N/R	N/R	N/R	N/S	1-1/16	1-3/16	1-3/8	1-7/16
N/S	N/R	N/R	N/R	N/R	N/R	N/R	N/S	1-1/16	1-3/16	1-3/8	1-7/16
N/S	N/R	N/R	N/R	N/R	N/R	N/R	N/S	1-1/16	1-3/16	1-3/8	1-7/16
108-83-2	108-89-2	108-83-2	108-89-2	N/A	N/S	N/S	N/S	1-1/4	1-5/16	1-11/16	1-11/16
108-83-2	108-89-2	108-83-2	108-89-2	N/A	N/S	N/S	N/S	1-1/4	1-5/16	1-11/16	1-11/16
108-83-2	108-89-2	108-83-2	108-89-2	N/A	N/S	N/S	N/S	1-1/4	1-5/16	1-11/16	1-11/16
108-83-2	108-89-2	108-83-2	108-89-2	N/A	N/S	N/S	N/S	1-3/8	1-3/8	1-11/16	1-11/16
108-83-2	108-89-2	108-83-2	108-89-2	N/A	N/S	N/S	N/S	1-3/8	1-3/8	1-11/16	1-11/16
108-83-2	108-89-2	108-83-2	108-89-2	N/A	N/S	N/S	N/S	1-3/8	1-3/8	1-11/16	1-11/16
108-83-2	108-91-2	108-90-2	108-90-2	108-27-2	N/S	N/S	N/S	1-1/4	1-5/16	1-9/16	1-9/16
108-92-2	108-91-2	108-90-2	108-90-2	108-27-2	N/S	N/A	N/S	1-13/32	1-13/64	1-3/8	2
N/S	N/R	N/R	N/R	N/R	N/R	N/R	N/S	1-1/16	1-3/16	1-3/8	1-7/16
N/S	N/R	N/R	N/R	N/R	N/R	N/R	N/S	1-1/16	1-3/16	1-3/8	1-7/16
N/S	N/R	N/R	N/R	N/R	N/R	N/R	N/S	1-1/16	1-3/16	1-3/8	1-7/16
N/S	N/R	N/R	N/R	N/R	N/R	N/R	N/S	1-1/16	1-3/16	1-3/8	1-7/16
108-83-2	108-91-2	108-90-2	108-90-2	108-27-2	134-101	134-105	N/S	1-1/4	1-5/16	1-9/16	1-9/16
108-83-2	108-91-2	108-90-2	108-90-2	108-27-2	N/S	134-105	N/S	1-1/4	1-5/16	1-9/16	1-9/16
108-83-2	108-91-2	108-90-2	108-90-2	108-27-2	N/S	134-105	N/S	1-1/4	1-5/16	1-9/16	1-9/16
N/S	N/R	N/R	N/R	N/R	N/R	N/R	N/S	1-1/16	1-3/16	1-3/8	1-7/16
N/S	N/R	N/R	N/R	N/R	N/R	N/R	N/S	1-1/16	1-3/16	1-3/8	1-7/16
N/S	N/R	N/R	N/R	N/R	N/R	N/R	N/S	1-1/16	1-3/16	1-3/8	1-7/16
108-83-2	108-91-2	108-90-2	108-90-2	108-27-2	N/S	134-105	N/S	1-1/4	1-5/16	1-9/16	1-9/16
108-83-2	108-91-2	108-90-2	108-90-2	108-27-2	N/S	134-105	N/S	1-1/4	1-5/16	1-9/16	1-9/16
108-83-2	108-91-2	108-90-2	108-90-2	108-27-2	N/S	134-105	N/S	1-1/4	1-5/16	1-9/16	1-9/16
108-83-2	108-91-2	108-90-2	108-90-2	108-27-2	N/A	134-105	N/A	1-1/16	1-1/16	1-7/16	1-7/16
108-92-2	108-91-2	108-90-2	108-90-2	108-27-2	134-110	N/A	N/S	1-13/64	1-13/32	1-3/8	2
108-92-2	108-91-2	108-90-2	108-90-2	108-27-2	134-110	N/A	N/S	1-13/64	1-13/32	1-3/8	
108-92-2	108-91-2	108-90-2	108-90-2	108-27-2	134-110	N/A	N/S	1-13/64	1-13/32	1-3/8	2
108-92-2	108-91-2	108-90-2	108-90-2	108-27-2	134-110	N/A	N/S	1-13/64	1-13/32	1-3/8	2
108-83-2	108-89-2	108-83-2	108-89-2	N/R	134-108	134-112	N/S	1.69	1.69	2	2
108-83-2	108-89-2	108-90-2	108-89-2	N/R	134-108	134-112	N/S	1-11/16	1-11/16	2	2
108-83-2	108-89-2	108-83-2	108-89-2	N/R	134-108	134-112	N/S	1-11/16	1-11/16	2	2
108-83-2	108-89-2	108-83-2	108-89-2	N/R	134-108	134-112	N/S	1-11/16	1-11/16	2	2
N/A	N/A	N/A	N/A	108-27-2	N/S	N/S	N/S	1-1/4	1-5/16	1-9/16	1-9/16
N/S	N/R	N/R	N/R	N/R	N/R	N/R	N/S	1-1/16	1-3/16	1-3/8	1-7/16
N/S	N/R	N/R	N/R	N/R	N/R	N/R	N/S	1-1/16	1-3/16	1-3/8	1-7/16
108-83-2	108-89-2	108-83-2	108-89-2	N/R	134-103	134-104	N/A	1-3/8	1-3/8	1-11/16	1-11/16

Carb Number	Carb Model Number	CFM	Renew Kit	Trick Kit	Needle & Seat	Primary Main Jet	Secondary Main Jet or plate	Primary Metering Block	Secondary Metering Block	Power Valve	Pump Nozzle Size, Primary	Secondary Nozzle Size or Spring Color
R8158	4360	450	37-1540	N/A	N/S	N/A	N/A	N/R	N/R	N/S	0.028	N/R
R8159	4160	450	703-33	N/A	6-511	122-59	34R9716-32	N/S	N/S	125-85	0.021	PLAIN
R8162	4150	850	37-485	37-933	6-504	122-80	122-80	N/A	N/S	125-65	0.031	0.031
R8181	4160	600	37-1536	37-933	6-504	122-80	122-80	N/S	134-39	125-65 (15)	0.031	0.031
R8203	4360	450	37-1540	N/A	N/S	N/A	N/A	N/R	N/R	N/S	0.028	N/R
R8204	4360	450	37-1540	N/A	N/S	N/A	N/A	N/R	N/R	N/S	0.028	N/R
R8206	4360	450	37-1540	N/A	N/S	N/A	N/A	N/R	N/R	N/S	0.028	N/R
R8207	4160	600	N/A	N/A	6-506	N/A	134-39	N/S	N/S	125-85	0.031	Plain
R8276	4175	650	37-1537	37-933	(16,17)	N/A	134-21	N/S	N/S	125-85	0.025	Black
R8302	4175	650	37-1537	37-933	(16,17)	N/A	134-21	N/S	N/S	125-85	0.025	Black
R8479	4360	450	37-1540	N/A	N/S	N/A	N/A	N/R	N/R	N/S	0.028	N/R
R8516	4360	450	37-1540	N/A	N/S	N/A	N/A	N/R	N/R	N/S	0.028	N/R
R8517	4360	450	37-1540	N/A	N/S	N/A	N/A	N/R	N/R	N/S	0.028	N/R
R8546	4175	650	37-1537	37-933	(16,17)	N/A	134-21	N/S	134-21	125-85	0.025	Black
R8572	4150	715	3-1184	N/A	6-504	122-72	122-84	N/S	N/S	125-85	0.026	YELLOW
R8642	4360	450	37-1540	N/A	N/S	N/A	N/A	N/R	N/R	N/S	0.028	N/R
R8677	4360	450	37-1540	N/A	N/S	N/A	N/A	N/R	N/R	N/S	0.028	N/R
R8679	4175	650	37-1537	37-933	(16,17)	N/A	34R9716-27	N/S	N/A	125-85	0.025	Plain
R8700	4175	650	37-1537	37-933	(16,17)	N/A	134-21	N/S	134-21	125-85	0.025	Black
R8771	4360	450	37-1540	N/A	N/S	N/A	N/A	N/R	N/R	N/S	0.028	N/R
R8804	4150	830	37-485	37-933	6-504	122-80	122-80	N/S	N/S	125-65 (B)	0.028	0.028
R8874	4360	450	37-1540	N/A	N/S	N/A	N/A	N/R	N/R	N/S	0.028	N/R
R8875	4360	450	3-1160	N/A	N/S	N/A	N/A	N/R	N/R	N/S	0.028	N/R
R8876	4360	450	N/A	N/A	N/S	N/A	N/A	N/R	N/R	N/S	0.028	N/R
R8877	4360	450	3-1160	N/A	N/S	N/A	N/A	N/R	N/R	N/S	0.028	N/R
R8879	4175	650	37-1537	37-933	(16,17)	N/A	134-21	N/S	134-21	125-65	0.025	Black
R8896	4500	1050	37-1539	37-933	6-518-2	122-90	122-90	N/A	N/A	122-55	0.035	0.035
R8896-1	4500	1050	37-1539	37-933	6-518-2	122-88	122-88	N/A	N/A	125-55	0.035	0.035
R8896-2	4500	1050	37-1539	37-933	6-518-2	122-88	122-88	134-70	134-70	125-55	0.035	0.035
R8914	4360	450	37-1540	N/A	N/S	N/A	N/A	N/R	N/R	N/S	0.028	N/R
R8958	4360	450	37-1540	N/A	N/S	124-195	N/A	N/R	N/R	N/S	0.028	N/R
R9002	4160	600	37-1536	37-933	6-506	122-632	134-37	N/S	134-37	125-208	0.031	Black
R9011	2300	500	3-474	N/A	6-504	122-75	N/S	N/S	N/S	125-50	0.028	N/S
R9013	4160	600	3-720	N/A	6-506	122-64	34R9716-44	N/S	N/S	125-65	0.031	BLACK
R9015	4160	750	3-720	N/A	6-504	122-76	34R9716-27	N/S	N/S	125-105	0.025	PLAIN
R9015-1	4160	750	3-720	N/A	6-504	122-76	34R9716-27	N/S	N/R	125-105	0.025	Plain
R9022	4150	800	3-485	N/A	6-504	122-72	122-87	N/S	N/S	125-65	0.031	0.031
R9023	4165	800	3-605	N/A	6-504	122-61	122-86	N/S	N/S	125-85 (15)	0.025	0.037
R9029	4150	715	3-1184	N/A	6-504	122-75	122-84	N/S	N/S	125-85	0.026	YELLOW
R9040	4160	600	37-119	37-933	N/S	122-661	N/S	N/S	N/S	N/A	0.031	Plain
R9088	4360	450	N/A	N/A	N/S	N/A	N/A	N/R	N/R	N/S	0.028	N/R
R9105	4360	450	3-1160	N/A	N/S	N/A	N/A	N/R	N/R	N/S	0.028	N/R
R9112	4360	450	37-1540	N/A	N/S	N/A	N/A	N/R	N/R	N/S	0.028	N/R
R9162	4360	450	37-1540	N/A	N/S	N/A	N/A	N/R	N/R	N/S	0.028	N/R
R9185	4360	450	37-1540	N/A	N/S	N/A	N/A	N/R	N/R	N/S	0.028	N/R
R9188	4150	780	37-1539	37-933	6-504	122-72	122-76	N/S	N/S	(12,21)	0.025	Plain
R9192	4360	450	37-1540	N/A	N/S	N/A	N/A	N/R	N/R	N/S	0.028	N/R
R9193	4360	450	37-1540	N/A	N/S	N/A	N/A	N/R	N/R	N/S	0.028	N/R
R9210	4160	600	37-1536	37-933	6-506	N/A	134-39	N/S	N/S	125-208	0.031	Black
R9219	4160	600	37-1536	37-933	6-506	N/A	134-39	N/S	134-39	125-208	0.031	Black
R9228	5200	280	N/A	N/A	N/S	N/A	N/A	N/R	N/R	N/S	0.023	N/R
R9254	4160	600	37-1536	37-933	6-506	N/A	134-39	N/S	N/S	N/A	0.031	Black
R9375	4500	1050	37-1539	37-933	6-518-2	122-86	122-86	N/A	N/A	PLUG	0.035	0.035

Primary Bowl Gasket	Primary Metering Block Gasket	Secondary Bowl Gasket	Secondary Metering Block Gasket	Secondary Metering Plate Gasket	Primary Fuel Bowl	Secondary Fuel Bowl	Throttle Body & Shaft Assembly	Venturii Diameter Primary	Venturii Diameter Secondary	Throttle Bore Diameter Primary	Throttle Bore Diameter Secondary
N/S	N/R	N/R	N/R	N/R	N/R	N/R	N/S	1-1/16	1-3/16	1-3/8	1-7/16
N/A	N/A	N/A	N/A	108-27-2	N/S	N/S	N/S	1-3/32	1-3/32	1-1/2	1-1/2
108-83-2	108-89-2	108-83-2	108-89-2	N/R	134-103	134-104	N/A	1-9/16	1-9/16	1-3/4	1-3/4
108-83-2	108-89-2	108-83-2	108-89-2	N/R	N/A	134-105	N/S	1-9/16	1-9/16	1-3/4	1-3/4
N/S	N/R	N/R	N/R	N/R	N/R	N/R	N/S	1-1/16	1-3/16	1-3/8	1-7/16
N/S	N/R	N/R	N/R	N/R	N/R	N/R	N/S	1-1/16	1-3/16	1-3/8	1-7/16
N/S	N/R	N/R	N/R	N/R	N/R	N/R	N/S	1-1/16	1-3/16	1-3/8	1-7/16
108-83-2	108-91-2	108-90-2	108-90-2	108-27-2	N/S	N/S	N/S	1-1/4	1-5/16	1-9/16	1-9/16
108-92-2	108-91-2	108-90-2	108-90-2	108-27-2	N/S	N/S	N/S	1-13/64	1-13/32	1-3/8	2
108-92-2	108-91-2	108-90-2	108-90-2	108-27-2	N/S	N/S	N/S	1-13/64	1-13/32	1-3/8	2
N/S	N/R	N/R	N/R	N/R	N/R	N/R	N/S	1-1/16	1-3/16	1-3/8	1-7/16
N/S	N/R	N/R	N/R	N/R	N/R	N/R	N/S	1-1/16	1-3/16	1-3/8	1-7/16
N/S	N/R	N/R	N/R	N/R	N/R	N/R	N/S	1-1/16	1-3/16	1-3/8	1-7/16
108-92-2	108-91-2	108-90-2	108-90-2	108-27-2	134-110	N/A	N/S	1-13/64	1-13/32	1-3/8	2
N/A	N/A	N/A	N/A	N/A	134-108	N/S	N/S	1-3/8	1-7/16	1-11/16	1-11/16
N/S	N/R	N/R	N/R	N/R	N/R	N/R	N/S	1-1/16	1-3/16	1-3/8	1-7/16
N/S	N/R	N/R	N/R	N/R	N/R	N/R	N/S	1-1/16	1-3/16	1-3/8	1-7/16
108-92-2	108-91-2	108-90-2	108-90-2	108-27-2	134-110	N/A	N/S	1-13/64	1-13/32	1-3/8	2
108-92-2	108-91-2	108-90-2	108-90-2	108-27-2	134-110	N/A	N/S	1-13/64	1-13/32	1-3/8	2
N/S	N/R	N/R	N/R	N/R	N/R	N/R	N/S	1-1/16	1-3/16	1-3/8	1-7/16
108-83-2	108-89-2	108-83-2	108-89-2	N/R	134-103	134-104	N/A	1-9/16	1-9/16	1-11/16	1-11/16
N/S	N/R	N/R	N/R	N/R	N/R	N/R	N/S	1-1/16	1-3/16	1-3/8	1-7/16
N/S	N/R	N/R	N/R	N/R	N/R	N/R	N/S	1-1/16	1-3/16	1-3/8	1-7/16
N/S	N/R	N/R	N/R	N/R	N/R	N/R	N/S	1-1/16	1-3/16	1-3/8	1-7/16
N/S	N/R	N/R	N/R	N/R	N/R	N/R	N/S	1-1/16	1-3/16	1-3/8	1-7/16
108-92-2	108-91-2	108-90-2	108-90-2	108-27-2	134-110	N/A	N/S	1-13/64	1-13/32	1-3/8	2
108-83-2	108-36-2	108-36-2	108-36-2	N/R	N/R	N/R	N/S	1.69	1.69	2	2
108-120	108-121	108-120	108-121	N/R	134-108	134-112	N/S	1-11/16	1-11/16	2	2
108-120	108-121	108-120	108-121	N/R	134-108	134-112	N/S	1-11/16	1-11/16	2	2
N/S	N/R	N/R	N/R	N/R	N/R	N/R	N/S	1-1/16	1-3/16	1-3/8	1-7/16
N/S	N/R	N/R	N/R	N/R	N/R	N/R	N/S	1-1/16	1-3/16	1-3/8	1-7/16
108-83-2	108-91-2	108-90-2	108-90-2	108-27-2	N/A	134-105	N/S	1-1/4	1-5/16	1-9/16	1-9/16
N/A	N/A	N/A	N/A	N/A	N/S	N/S	N/S	1-3/8	N/R	1-11/16	N/R
N/A	N/A	N/A	N/A	108-27-2	N/S	N/S	N/S	1-1/4	1-5/16	1-9/16	1-9/16
N/A	N/A	N/A	N/A	108-27-2	N/S	N/S	N/S	1-3/8	1-7/16	1-11/16	1-11/16
N/A	N/A	N/A	N/A	108-27-2	N/S	N/S	N/S	1-3/8	1-7/16	1-11/16	1-11/16
N/A	N/A	N/A	N/A	N/R	N/S	N/S	N/S	1-3/8	1-7/16	1-11/16	1-11/16
N/A	N/A	N/A	N/A	N/A	N/S	N/S	N/S	1-5/32	1-23/32	1-3/8	2
N/A	N/A	N/A	N/A	N/A	N/S	N/S	N/S	1-3/8	1-7/16	1-11/16	1-11/16
108-83-2	108-91-2	108-90-2	108-90-2	108-27-2	N/S	134-105	N/S	1-1/4	1-5/16	1-9/16	1-9/16
N/S	N/R	N/R	N/R	N/R	N/R	N/R	N/S	1-1/16	1-3/16	1-3/8	1-7/16
N/S	N/R	N/R	N/R	N/R	N/R	N/R	N/S	1-1/16	1-3/16	1-3/8	1-7/16
N/S	N/R	N/R	N/R	N/R	N/R	N/R	N/S	1-1/16	1-3/16	1-3/8	1-7/16
N/S	N/R	N/R	N/R	N/R	N/R	N/R	N/S	1-1/16	1-3/16	1-3/8	1-7/16
N/S	N/R	N/R	N/R	N/R	N/R	N/R	N/S	1-1/16	1-3/16	1-3/8	1-7/16
108-83-2	108-89-2	108-83-2	108-89-2	N/R	N/S	N/S	N/S	1-3/8	1-7/16	1-11/16	1-11/16
N/S	N/R	N/R	N/R	N/R	N/R	N/R	N/S	1-1/16	1-3/16	1-3/8	1-7/16
N/S	N/R	N/R	N/R	N/R	N/R	N/R	N/S	1-1/16	1-3/16	1-3/8	1-7/16
108-83-2	108-91-2	108-90-2	108-90-2	108-27-2	N/S	N/S	N/S	1-1/4	1-5/16	1-9/16	1-9/16
108-83-2	108-91-2	108-90-2	108-90-2	108-27-2	N/S	134-105	N/S	1-1/4	1-5/16	1-9/16	1-9/16
N/R	N/R	N/R	N/R	N/R	N/R	N/R	N/R	1-1/25	1-1/16	1-7/25	1-7/16
108-83-2	108-91-2	108-90-2	108-90-2	108-27-2	N/S	N/S	N/S	1-1/4	1-5/16	1-9/16	1-9/16
108-83-2	108-36-2	108-36-2	108-36-2	N/R	N/S	N/S	N/S	1.69	1.69	2	2

Carb Number	Carb Model Number	CFM	Renew Kit	Trick Kit	Needle & Seat	Primary Main Jet	Secondary Main Jet or plate	Primary Metering Block	Secondary Metering Block	Power Valve	Pump Nozzle Size, Primary	Secondary Nozzle Size or Spring Color
R9375-1	4500	1050	37-1539	37-933	6-518-2	122-86	122-86	N/A	N/A	N/R	0.035	0.035
R9375-2	4500	1050	37-1539	37-933	6-518-2	122-86	122-86	N/A	N/A	N/R	0.035	0.035
R9377	4500	1150	37-1539	37-933	6-518-2	122-96	122-96	N/S	N/S	PLUG	0.035	0.035
R9377-1	4500	1150	37-1539	37-933	6-518-2	122-92	122-92	N/A	N/A	N/R	0.035	0.037
R9377-2	4500	1150	37-1539	37-933	6-518-2	122-92	122-92	N/A	N/A	N/R	0.035	0.037
R9379	4150	750	37-485	37-933	6-504	122-68	122-81	134-155	N/A	125-65	0.028	0.031
R9379-1	4150	750	37-485	37-933	6-504	122-68	122-81	134-155	N/A	125-65	0.028	0.031
R9380	4150	850	37-485	37-933	6-504	122-78	122-78	N/A	N/S	125-65 (15)	0.031	0.031
R9380-1	4150	850	37-485	37-933	6-504	122-78	122-78	N/A	N/S	125-65 (15)	0.031	0.031
R9381	4150	830	37-485	37-933	6-504	122-78	122-78	N/A	N/S	125-65 (15)	0.028	0.028
R9392	4160	600	703-33	N/A	6-511	122-66	N/S	N/S	N/S	125-25	0.025	PURPLE
R9393	4160	450	703-28	N/A	6-511	122-59	N/S	N/S	N/S	125-85	0.021	PLAIN
R9394	4160	650	703-28	N/A	6-511	122-73	N/S	N/S	N/S	125-65	0.026	YELLOW
R9399	4160	650	703-28	N/A	6-511	122-73	N/S	N/S	N/S	125-65	0.04	WHITE
R9429	5200	280	N/A	N/A	N/S	N/A	124-231	N/R	N/R	N/S	0.023	N/R
R9441	5200	280	N/A	N/A	N/S	N/A	124-231	N/R	N/R	N/S	0.023	N/R
R9444	5200	280	N/A	N/A	N/S	N/A	124-231	N/R	N/R	N/S	0.023	N/R
R9446	5200	280	N/A	N/A	N/S	N/A	124-231	N/R	N/R	N/S	0.023	N/R
R9545	5200	280	N/A	N/A	N/S	N/A	124-231	N/R	N/R	N/S	0.023	N/R
R9626	4160	600	3-1415	N/A	6-506	N/A	134-39	N/S	N/S	125-206	0.031	Black
R9644	6520	280	N/A	N/A	N/A	N/A	N/A	N/R	N/R	N/A	0.02	N/R
R9645	4150	750	37-1539	37-933	6-515-2	122-80	122-80	N/A	N/A	125-165 (15)	0.045	0.045
R9646	4150	850	37-1539	37-933	6-515-2	122-92	122-92	N/A	N/A	125-165 (15)	0.045	0.045
R9647	2300	500	37-1536	37-933	6-515-2	122-81	N/R	N/A	N/R	125-145	0.04	N/R
R9655	6520	280	N/A	N/A	N/A	N/A	N/A	N/R	N/R	N/A	0.02	N/R
R9659	6520	280	N/A	N/A	N/A	N/A	N/A	N/R	N/R	N/A	0.02	N/R
R9678	4360	450	3-1160	N/A	N/S	N/A	N/A	N/R	N/R	N/S	0.028	N/R
R9681	5200	280	N/A	N/A	N/S	N/A	N/A	N/R	N/R	N/S	0.023	N/R
R9682	6520	280	N/A	N/A	N/A	N/A	N/A	N/R	N/R	N/A	0.02	N/R
R9688	5200	280	N/A	N/A	N/S	N/A	N/A	N/R	N/R	N/S	0.023	N/R
R9689	5200	280	N/A	N/A	N/S	N/A	N/A	N/R	N/R	N/S	0.023	N/R
R9694	4360	450	37-1540	N/A	N/S	N/A	N/A	N/R	N/R	N/S	0.028	N/R
R9767	5200	280	N/A	N/A	N/S	N/A	N/A	N/R	N/R	N/S	0.023	N/R
R9776	4160	450	37-1536	37-933	6-506	N/A	N/A	N/S	N/A	125-85	0.031	N/R
R9777	4360	450	37-1540	N/A	N/S	N/A	N/A	N/R	N/R	N/S	0.028	N/R
R9781	5200	280	N/A	N/A	N/S	N/A	N/A	N/R	N/R	N/S	0.023	N/R
R9810	6520	280	N/A	N/A	N/A	N/A	N/A	N/R	N/R	N/A	0.02	N/R
R9811	6520	280	N/A	N/A	N/A	N/A	N/A	N/R	N/R	N/A	0.02	N/R
R9834	4160	600	37-720	37-933	6-506	N/A	134-39	N/S	134-39	125-65	0.031	Black
R9834-1	4160	600	37-720	37-933	6-506	N/A	134-39	N/S	134-39	125-65	0.031	Black
R9834-2	4160	600	37-720	37-933	6-506	122-68	134-39	N/S	134-39	125-65	0.031	Black
R9834-3	4160	600	37-720	37-933	6-506	122-68	134-39	N/S	134-39	125-65	0.031	Black
R9864	5200	280	N/A	N/A	N/S	N/A	N/A	N/R	N/R	N/S	0.023	N/R
R9875	4360	450	N/A	N/A	N/S	N/A	N/A	N/R	N/R	N/S	0.028	N/R
R9895	4175	650	37-1537	37-933	(16,17)	N/A	134-21	N/S	134-21	125-206	0.037	Black
R9895-1	4175	650	37-1537	37-933	(16,17)	N/A	134-21	N/S	134-21	125-206	0.037	Black
R9896	6510	280	N/A	N/A	N/A	N/A	N/A	N/R	N/R	N/A	0.02	N/R
R9899	5200	280	N/A	N/A	N/S	N/A	N/A	N/R	N/R	N/S	0.023	N/R
R9923	4175	650	37-1537	37-933	(16,17)	N/A	N/S	N/S	N/S	N/A	0.025	Black
R9925	5200	280	N/A	N/A	N/S	N/A	N/A	N/R	N/R	N/S	0.023	N/R
R9931	4360	450	37-1540	N/A	N/S	N/A	N/A	N/R	N/R	N/S	0.028	N/R
R9932	5200	280	N/A	N/A	N/S	N/A	N/A	N/R	N/R	N/S	0.023	N/R
R9935	4360	450	37-1540	N/A	N/S	N/A	N/A	N/R	N/R	N/S	0.028	N/R

Primary Bowl Gasket	Primary Metering Block Gasket	Secondary Bowl Gasket	Secondary Metering Block Gasket	Secondary Metering Plate Gasket	Primary Fuel Bowl	Secondary Fuel Bowl	Throttle Body & Shaft Assembly	Venturii Diameter Primary	Venturii Diameter Secondary	Throttle Bore Diameter Primary	Throttle Bore Diameter Secondary
108-120	108-121	108-120	108-121	N/R	134-108	134-112	N/S	1-11/16	1-11/16	2	2
108-120	108-121	108-120	108-121	N/R	134-108	134-112	N/S	1-11/16	1-11/16	2	2
108-83-2	108-36-2	108-36-2	108-36-2	N/R	N/R	N/R	N/S	1.83	1.83	2	2
108-120	108-121	108-120	108-121	N/R	134-108	134-112	N/S	1-13/16	1-13/16	2	2
108-120	108-121	108-120	108-121	N/R	134-108	134-112	N/S	1-13/16	1-13/16	2	2
108-83-2	108-89-2	108-83-2	108-89-2	N/R	134-103	134-104	N/A	1-3/8	1-3/8	1-11/16	1-11/16
108-83-2	108-89-2	108-83-2	108-89-2	N/R	134-103	134-104	N/A	1-3/8	1-3/8	1-11/16	1-11/16
108-83-2	108-89-2	108-83-2	108-89-2	N/R	134-103	134-104	N/A	1-9/16	1-9/16	1-3/4	1-3/4
108-83-2	108-89-2	108-83-2	108-89-2	N/R	134-103	134-104	N/A	1-9/16	1-9/16	1-3/4	1-3/4
108-83-2	108-89-2	108-83-2	108-89-2	N/R	134-103	134-104	N/A	1-9/16	1-9/16	1-11/16	1-11/16
N/A	N/A	N/A	N/A	108-13-2	N/S	N/S	N/S	1-1/4	1-5/16	1-9/16	1-9/16
N/A	N/A	N/A	N/A	108-13-2	N/S	N/S	N/S	1-3/32	1-3/32	1-1/2	1-1/2
N/A	N/A	N/A	N/A	108-13-2	N/S	N/S	N/S	1-1/4	1-5/16	1-9/16	1-9/16
N/A	N/A	N/A	N/A	108-13-2	N/S	N/S	N/S	1-1/4	1-5/16	1-9/16	1-9/16
N/R	N/R	N/R	N/R	N/R	N/R	N/R	N/R	1-1/25	1-1/16	1-7/25	1-7/16
N/R	N/R	N/R	N/R	N/R	N/R	N/R	N/R	1-1/25	1-1/16	1-7/25	1-7/16
N/R	N/R	N/R	N/R	N/R	N/R	N/R	N/R	1-1/25	1-1/16	1-7/25	1-7/16
N/R	N/R	N/R	N/R	N/R	N/R	N/R	N/R	1-1/25	1-1/16	1-7/25	1-7/16
N/R	N/R	N/R	N/R	N/R	N/R	N/R	N/R	1-1/25	1-1/16	1-7/25	1-7/16
108-83-2	108-91-2	108-90-2	108-90-2	108-27-2	N/S	N/S	N/S	1-1/4	1-5/16	1-9/16	1-9/16
N/R	N/R	N/R	N/R	N/R	N/R	N/R	N/R	1-1/25	1-1/16	1-7/25	1-7/16
108-83-2	108-89-2	108-83-2	108-89-2	N/R	134-103	134-104	N/A	1-3/8	1-3/8	1-11/16	1-11/16
108-83-2	108-89-2	108-83-2	108-89-2	N/R	134-103	134-104	N/S	1-9/16	1-9/16	1-3/4	1-3/4
108-83-2	108-89-2	N/R	N/R	N/R	134-103	N/S	112-2	1-3/8	N/R	1-11/16	N/R
N/R	N/R	N/R	N/R	N/R	N/R	N/R	N/R	1-1/25	1-1/16	1-7/25	1-7/16
N/R	N/R	N/R	N/R	N/R	N/R	N/R	N/R	1-1/25	1-1/16	1-7/25	1-7/16
N/S	N/R	N/R	N/R	N/R	N/R	N/R	N/S	1-1/16	1-3/16	1-3/8	1-7/16
N/R	N/R	N/R	N/R	N/R	N/R	N/R	N/R	1-1/25	1-1/16	1-7/25	1-7/16
N/R	N/R	N/R	N/R	N/R	N/R	N/R	N/R	1-1/25	1-1/16	1-7/25	1-7/16
N/R	N/R	N/R	N/R	N/R	N/R	N/R	N/R	1-1/25	1-1/16	1-7/25	1-7/16
N/R	N/R	N/R	N/R	N/R	N/R	N/R	N/R	1-1/25	1-1/16	1-7/25	1-7/16
N/S	N/R	N/R	N/R	N/R	N/R	N/R	N/R	1-1/16	1-3/16	1-3/8	1-7/16
N/R	N/R	N/R	N/R	N/R	N/R	N/R	N/R	1-1/25	1-1/16	1-7/25	1-7/16
108-83-2	108-89-2	108-90-2	108-90-2	108-27-2	134-101	134-105	N/A	1-3/32	1-3/32	1-1/2	1-1/2
N/S	N/R	N/R	N/R	N/R	N/R	N/R	N/R	1-1/16	1-3/16	1-3/8	1-7/16
N/R	N/R	N/R	N/R	N/R	N/R	N/R	N/R	1-1/25	1-1/16	1-7/25	1-7/16
N/R	N/R	N/R	N/R	N/R	N/R	N/R	N/R	1-1/25	1-1/16	1-7/25	1-7/16
N/R	N/R	N/R	N/R	N/R	N/R	N/R	N/R	1-1/25	1-1/16	1-7/25	1-7/16
108-83-2	108-91-2	108-90-2	108-90-2	108-27-2	N/A	134-105	N/S	1-1/4	1-5/16	1-9/16	1-9/16
108-83-2	108-91-2	108-90-2	108-90-2	108-27-2	N/A	134-105	N/S	1-1/4	1-5/16	1-9/16	1-9/16
108-83-2	108-91-2	108-90-2	108-90-2	108-27-2	N/A	134-105	N/S	1-1/4	1-5/16	1-9/16	1-9/16
108-83-2	108-91-2	108-90-2	108-90-2	108-27-2	N/A	134-105	N/S	1-1/4	1-5/16	1-9/16	1-9/16
N/R	N/R	N/R	N/R	N/R	N/R	N/R	N/R	1-1/25	1-1/16	1-7/25	1-7/16
N/S	N/R	N/R	N/R	N/R	N/R	N/R	N/R	1-1/16	1-3/16	1-3/8	1-7/16
108-92-2	108-91-2	108-90-2	108-90-2	108-27-2	134-110	N/A	N/A	1-13/64	1-13/32	1-3/8	2
108-92-2	108-91-2	108-90-2	108-90-2	108-27-2	134-110	N/A	N/A	1-13/64	1-13/32	1-3/8	2
N/R	N/R	N/R	N/R	N/R	N/R	N/R	N/R	1-1/25	1-1/16	1-7/25	1-7/16
N/R	N/R	N/R	N/R	N/R	N/R	N/R	N/R	1-1/25	1-1/16	1-7/25	1-7/16
108-92-2	108-91-2	108-90-2	108-90-2	108-27-2	N/S	N/S	N/S	1-13/64	1-13/32	1-3/8	2
N/R	N/R	N/R	N/R	N/R	N/R	N/R	N/R	1-1/25	1-1/16	1-7/25	1-7/16
N/S	N/R	N/R	N/R	N/R	N/R	N/R	N/S	1-1/16	1-3/16	1-3/8	1-7/16
N/R	N/R	N/R	N/R	N/R	N/R	N/R	N/R	1-1/25	1-1/16	1-7/25	1-7/16
N/S	N/R	N/R	N/R	N/R	N/R	N/R	N/S	1-1/16	1-3/16	1-3/8	1-7/16

Carb Number	Carb Model Number	CFM	Renew Kit	Trick Kit	Needle & Seat	Primary Main Jet	Secondary Main Jet or plate	Primary Metering Block	Secondary Metering Block	Power Valve	Pump Nozzle Size, Primary	Secondary Nozzle Size or Spring Color
R9948	4175	650	37-1537	37-933	(16,17)	N/A	N/S	N/S	N/S	N/A	0.025	Black
R9973	4360	450	37-1540	N/A	N/S	N/A	N/A	N/R	N/R	N/S	0.028	N/R
R9976	4175	650	37-1537	37-933	(16,17)	N/A	N/S	N/S	N/S	N/A	0.025	Black
R50399	4160	650	703-28	N/A	6-511	122-73	N/S	N/S	N/R	125-65	0.04	White
R50399-1	4160	650	703-28	N/A	N/S	122-73	N/S	N/S	N/R	125-65	0.04	White
R50405	4160	650	703-28	N/A	N/A	122-74	N/S	N/S	N/S	125-65	0.04	PINK
R50417	2300	300	703-30	N/A	6-511	122-60	N/S	N/S	N/S	125-50	0.028	N/S
R50418	4160	450	703-28	N/A	6-511	122-59	N/S	N/S	N/S	125-85	0.021	PLAIN
R50419	4160	600	703-29	N/A	6-511	122-66	N/S	N/S	N/S	125-25	0.025	PURPLE
R50419-1	4160	600	703-29	N/A	6-511	122-65	N/S	N/S	N/S	125-25	0.025	PURPLE
R50461	2300	300	703-30	N/A	6-511	122-60	N/S	N/S	N/S	125-50	0.028	N/S
R50462	4160	450	703-28	N/A	N/A	122-59	N/S	N/S	N/S	125-85	0.021	PLAIN
R50463	4160	600	703-29	N/A	N/A	122-65	N/S	N/S	N/S	125-25	0.025	PURPLE
R50464	4160	750	703-33	N/A	6-511	122-74	N/S	N/S	N/S	125-65	0.04	PINK
R50467	2300	300	703-30	N/A	6-511	122-61	N/S	N/S	N/S	125-50	0.028	N/S
R50468	4160	450	703-28	N/A	6-511	122-59	N/S	N/S	N/S	125-85	0.021	PLAIN
R50469	4160	600	703-47	N/A	6-511	122-65	N/S	N/S	N/S	125-25	0.025	BROWN
R50470	4160	650	703-33	N/A	N/A	122-74	N/S	N/S	N/S	125-65	0.04	PINK
R50483	4010	600	703-53	N/A	6-504	122-69	122-76	N/S	N/S	125-65	0.026	PINK
R50483-1	4010	600	703-53	N/A	6-504	122-69	122-77	N/S	N/S	125-65	0.035	PINK
R75007-1	4160	600	N/A	N/A	6-511	122-63	34R9716-60	34R11954A	N/S	125-25	0.037	RED
R75009	4160	600	703-29	N/A	6-511	122-66	34R6153-59	34R11962A	N/S	125-65	0.031	PINK
R75009-1	4160	600	703-29	N/A	6-511	122-66	34R6153-59	34R11962A	N/S	125-65	0.031	PLAIN
R75010	4500	1150	37-1539	37-933	6-518-2	122-92	122-92	34R11972-3A	34R11972-3A	N/R	0.035	0.037
R75011	4500	1250	37-1539	37-933	6-518-2	122-97	122-97	34R11972-5A	34R11972-5A	N/R	0.035	0.037
R75021	4150	600	703-45	N/A	6-504	122-70	122-76	34R12019A	34R6497-3AM	125-25	0.031	RED
R75021-1	4150	600	703-45	N/A	6-504	122-68	122-66	34R12019A	34R6497-3AM	125-25	0.031	WHITE
R76650AA	AL 4150	650	37-485	37-933	6-504	122-63	122-73	N/S	N/S	125-65	0.028	0.031
R76750AA	AL 4150	750	37-485	37-933	6-504	122-72	122-80	N/S	N/S	125-65	0.028	0.031
R80054	5200	280	N/A	N/A	N/S	N/A	N/A	N/R	N/R	N/S	0.023	N/R
R80055	5200	280	N/A	N/A	N/S	N/A	N/A	N/R	N/R	N/S	0.023	N/R
R80056	5200	280	N/A	N/A	N/S	N/A	N/A	N/R	N/R	N/S	0.023	N/R
R80057	5200	280	N/A	N/A	N/S	N/A	N/A	N/R	N/R	N/S	0.023	N/R
R80073	4175	650	N/A	N/A	(16,17)	N/A	N/S	N/S	N/S	N/A	0.037	Black
R80086	4360	450	N/A	N/A	N/S	N/A	N/A	N/R	N/R	N/S	0.028	N/R
R80095	2305	500	37-1536	37-933	6-504	122-55	122-73	N/S	N/R	125-85	0.035	0.028
R80098	4180	600	37-1536	37-933	6-506	N/A	N/S	N/S	N/S	N/A	0.028	Purple
R80099	4180	600	37-1536	37-933	6-506	N/A	N/S	N/S	N/S	N/A	0.028	Purple
R80111	4180	600	37-1536	37-933	6-517	N/A	N/S	N/S	N/S	N/A	0.028	Orange
R80112	4180	600	37-1536	37-933	6-517	N/A	N/S	N/S	N/S	N/A	0.028	Orange
R80120	2305	350	37-1536	37-933	6-504	122-52	122-65	N/S	N/R	125-85	0.035	0.028
R80128	4175	650	37-1537	37-933	6-510	N/A	N/S	N/S	N/S	N/A	0.031	White
R80133	4180	600	37-1536	37-933	6-506	N/A	N/S	N/S	N/S	N/A	0.028	Orange
R80134	4180	600	37-1536	37-933	6-506	N/A	N/S	N/S	N/S	N/S	0.028	Brown
R80135	4180	600	37-1536	37-933	6-506	N/A	N/S	N/S	N/S	N/S	0.028	Brown
R80136	4180	600	37-1536	37-933	6-506	N/A	N/S	N/S	N/S	N/S	0.028	Brown
R80137	4180	600	37-1536	37-933	6-506	N/A	N/S	N/S	N/S	N/S	0.028	Brown
R80139	4175	650	37-1537	37-933	6-510	N/A	134-21	N/S	N/S	N/S	0.037	Black
R80140	4175	650	N/A	N/A	6-510	N/A	N/S	N/S	N/S	N/A	0.037	Black
R80145	4150	600	37-1539	37-933	6-504	122-68	122-70	N/S	N/S	125-65	0.031	Plain

Primary Bowl Gasket	Primary Metering Block Gasket	Secondary Bowl Gasket	Secondary Metering Block Gasket	Secondary Metering Plate Gasket	Primary Fuel Bowl	Secondary Fuel Bowl	Throttle Body & Shaft Assembly	Venturii Diameter Primary	Venturii Diameter Secondary	Throttle Bore Diameter Primary	Throttle Bore Diameter Secondary
108-92-2	108-91-2	108-90-2	108-90-2	108-27-2	N/S	N/S	N/S	1-13/64	1-13/32	1-3/8	2
N/S	N/R	N/R	N/R	N/R	N/R	N/R	N/S	1-1/16	1-3/16	1-3/8	1-7/16
108-92-2	108-91-2	108-90-2	108-90-2	108-27-2	N/S	N/A	N/S	1-13/64	1-13/32	1-3/8	2
108-83-2	108-91-2	108-90-2	108-90-2	108-13-2	N/S	N/S	N/S	1-1/4	1-5/16	1-9/16	1-9/16
108-83-2	108-91-2	108-90-2	108-90-2	108-13-2	N/S	N/S	N/S	1-1/4	1-5/16	1-9/16	1-9/16
N/A	N/A	N/A	N/A	108-13-2	N/S	N/S	N/S	1-1/4	1-5/16	1-9/16	1-9/16
N/A	N/A	N/A	N/A	N/A	N/S	N/S	N/S	1-3/16	N/R	1-1/2	N/R
N/A	N/A	N/A	N/A	108-13-2	N/S	N/S	N/S	1-3/32	1-3/32	1-1/2	1-1/2
N/A	N/A	N/A	N/A	108-13-2	N/S	N/S	N/S	1-1/4	1-5/16	1-9/16	1-9/16
N/A	N/A	N/A	N/A	108-13-2	N/S	N/S	N/S	1-1/4	1-5/16	1-9/16	1-9/16
N/A	N/A	N/A	N/A	N/A	N/S	N/S	N/S	1-3/16	N/R	1-1/2	N/R
N/A	N/A	N/A	N/A	108-13-2	N/S	N/S	N/S	1-3/32	1-3/32	1-1/2	1-1/2
N/A	N/A	N/A	N/A	108-13-2	N/S	N/S	N/S	1-1/4	1-5/16	1-9/16	1-9/16
N/A	N/A	N/A	N/A	108-13-2	N/S	N/S	N/S	1-3/8	1-7/16	1-11/16	1-11/16
N/A	N/A	N/A	N/A	N/A	N/S	N/S	N/S	1-3/16	N/R	1-1/2	N/R
N/A	N/A	N/A	N/A	108-13-2	N/S	N/S	N/S	1-3/32	1-3/32	1-1/2	1-1/2
N/A	N/A	N/A	N/A	108-13-2	N/S	N/S	N/S	1-1/4	1/516	1-9/16	1-9/16
N/A	N/A	N/A	N/A	108-13-2	N/S	N/S	N/S	1-1/4	1-5/16	1-9/16	1-9/16
-3	-3	-3	-3	-3	N/S	N/S	N/S	1-1/4	1-1/4	1-11/16	1-11/16
-3	-3	-3	-3	-3	N/S	N/S	N/S	1-1/4	1-1/4	1-11/16	1-11/16
N/A	N/A	N/A	N/A	108-27-2	34R7800-3AM	34R5972-3AM	12R11257A	1-1/4	1-5/16	1-9/16	1-9/16
N/A	N/A	N/A	N/A	108-13-2	34R10918A	34R5972-3AM	12R10830A	1-1/4	1-5/16	1-9/16	1-9/16
N/A	N/A	N/A	N/A	108-13-2	34R10918A	34R5972-3AM	12R10830A	1-1/4	1-5/16	1-9/16	1-9/16
108-120	108-121	108-120	108-121	N/A	134-108	134-112	N/S	1-13/16	1-13/16	2	2
108-120	108-121	108-120	108-121	N/A	134-108	134-112	N/S	1-13/16	1-13/16	2	2
N/A	N/A	N/A	N/A	N/A	34R11341-1	34R11340-1	12R11315A	1-1/4	1-5/16	1-9/16	1-9/16
N/A	N/A	N/A	N/A	N/A	34R11341-1	34R11340-1	12R11315A	1-1/4	1-5/16	1-9/16	1-9/16
108-83-2	108-89-2	108-83-2	108-89-2	N/A	N/S	N/S	N/S	1.25	1.3125	1.6875	1.6875
108-83-2	108-89-2	108-83-2	108-89-2	N/A	N/S	N/S	N/S	1.375	1.375	1.6875	1.6875
N/S	N/R	N/R	N/R	N/R	N/R	N/R	N/R	1-1/25	1-1/16	1-7/25	1-7/16
N/S	N/R	N/R	N/R	N/R	N/R	N/R	N/R	1-1/25	1-1/16	1-7/25	1-7/16
N/S	N/R	N/R	N/R	N/R	N/R	N/R	N/R	1-1/25	1-1/16	1-7/25	1-7/16
N/S	N/R	N/R	N/R	N/R	N/R	N/R	N/R	1-1/25	1-1/16	1-7/25	1-7/16
N/A	108-91-2	108-90-2	108-90-2	108-27-2	N/S	N/S	N/S	1-13/64	1-13/32	1-3/8	2
N/S	N/R	N/R	N/R	N/R	N/R	N/R	N/R	1-1/16	1-3/16	1-3/8	1-7/16
108-83-2	108-89-2	N/R	N/R	N/R	N/R	N/R	N/R	1-3/8	1-3/8	1-11/16	1-11/16
108-56-2	108-55-2	108-90-2	108-90-2	108-13-2	N/S	N/S	N/S	1-1/4	1-5/16	1-9/16	1-9/16
108-56-2	108-55-2	108-90-2	108-90-2	108-13-2	N/S	N/S	N/S	1-1/4	1-5/16	1-9/16	1-9/16
108-56-2	108-55-2	108-90-2	108-90-2	108-13-2	N/S	N/S	N/S	1-1/4	1-5/16	1-9/16	1-9/16
108-56-2	108-55-2	108-90-2	108-90-2	108-13-2	N/S	N/S	N/S	1-1/4	1-5/16	1-9/16	1-9/16
108-83-2	108-89-2	N/R	N/R	N/R	N/S	N/S	N/S	1-3/16	1-3/16	1-11/16	1-11/16
108-92-2	108-91-2	108-90-2	108-90-2	108-27-2	N/S	N/S	N/S	1-13/32	1-13/64	1-3/8	2
108-56-2	108-55-2	108-90-2	108-90-2	108-13-2	N/S	N/S	N/S	1-1/4	1-5/16	1-9/16	1-9/16
108-56-2	108-55-2	108-90-2	108-90-2	108-13-2	N/S	N/S	N/S	1-1/4	1-5/16	1-9/16	1-9/16
108-56-2	108-55-2	108-90-2	108-90-2	108-13-2	N/S	N/S	N/S	1-1/4	1-5/16	1-9/16	1-9/16
108-56-2	108-55-2	108-90-2	108-90-2	108-13-2	N/S	N/S	N/S	1-1/4	1-5/16	1-9/16	1-9/16
108-56-2	108-55-2	108-90-2	108-90-2	108-13-2	N/S	N/S	N/S	1-1/4	1-5/16	1-9/16	1-9/16
108-92-2	108-91-2	108-90-2	108-90-2	108-27-2	N/S	N/S	N/S	1-13/32	1-13/64	1-3/8	2
N/A	108-91-2	108-90-2	108-90-2	108-27-2	N/S	N/S	N/S	1-13/32	1-13/64	1-3/8	2
108-83-2	108-91-2	108-83-2	108-89-2	N/R	N/S	N/A	N/S	1-1/4	1-5/16	1-9/16	1-9/16

Carb Number	Carb Model Number	CFM	Renew Kit	Trick Kit	Needle & Seat	Primary Main Jet	Secondary Main Jet or plate	Primary Metering Block	Secondary Metering Block	Power Valve	Pump Nozzle Size, Primary	Secondary Nozzle Size or Spring Color
R80155	4175	650	37-1537	37-933	6-510	N/A	134-21	N/S	N/S	N/S	0.037	Black
R80159	4150	715	3-1184	N/A	6-504	122-74	122-85	N/S	N/S	125-85	0.026	YELLOW
R80163	4180	600	37-1536	37-933	6-506	N/A	N/S	N/S	N/S	N/S	0.028	Purple
R80164	4180	600	37-1536	37-933	6-506	N/A	N/S	N/S	N/S	N/S	0.028	Brown
R80165	4180	600	37-1536	37-933	6-506	N/A	N/S	N/S	N/S	N/S	0.028	Pink
R80166	4180	600	37-1536	37-933	6-506	N/A	N/S	N/S	N/S	N/S	0.028	Pink
R80169	4175	650	37-1537	37-933	6-510	N/A	N/A	N/S	N/S	N/A	0.025	Black
R80180	4150	850	703-46	N/A	6-504	122-92	122-92	N/S	N/S	125-65 (15)	0.028	YELLOW
R80186	4500	750	37-1539	37-933	6-518-2	122-74	122-74	N/S	N/S	125-85	0.028	0.028
R80186-1	4500	750	37-1539	37-933	6-518-2	122-70	122-70	N/S	N/S	125-65 (15)	0.028	0.035
R80186-2	4500	750	37-1539	37-933	6-518-2	122-70	122-70	N/S	N/S	125-65 (15)	0.028	0.035
R80262	4160	650	703-28	N/A	N/A	122-74	N/S	N/S	N/S	125-65	0.04	PINK
R80263	2300	300	703-30	N/A	N/A	122-60	N/S	N/S	N/S	125-50	0.028	N/S
R80264	4160	450	703-28	N/A	N/A	122-59	N/S	N/S	N/S	125-85	0.021	PLAIN
R80265	4160	600	703-29	N/A	N/A	122-65	N/S	N/S	N/S	125-25	0.025	PURPLE
R80309	4150	715	703-45	N/A	6-504	122-72	-1	N/S	N/S	125-25	0.031	YELLOW
R80310	4175	650	703-34	N/A	6-511	122-61	N/S	N/S	N/S	125-50	0.04	RED
R80311	4150	850	703-35	N/A	6-504	122-84	122-88	N/S	N/S	125-65 (22)	0.04	YELLOW
R80312	2300	350	703-36	N/A	6-511	122-70	N/S	N/S	N/S	125-25	0.028	N/S
R80313	2300	350	703-41	N/A	6-506	122-62	N/S	N/S	N/S	125-25	0.031	N/S
R80315	4160	600	703-29	N/A	6-511	122-67	N/S	N/S	N/S	125-25	0.025	YELLOW
R80316	2300	500	703-41	N/A	6-506	122-75	N/S	N/S	N/S	125-25	0.028	N/S
R80318-1	4160	600	703-33	N/A	N/S	122-74	N/S	N/S	N/S	125-65	0.04	Pink
R80319-1	4160	600	703-47	N/A	6-511	122-65	N/S	N/S	N/S	125-25	0.025	Brown
R80319-2	4160	600	703-47	N/A	6-511	122-65	N/S	N/S	N/S	125-25	0.025	Brown
R80319-3	4160	600	703-47	N/A	6-511	122-65	N/S	N/S	N/S	125-25	0.025	Brown
R80319-4	4160	600	703-47	N/A	6-511	122-65	N/S	N/S	N/S	125-25	0.025	Brown
R80320-1	2300	350	703-30	N/A	N/S	122-61	N/R	N/S	N/R	125-50	0.028	N/R
R80320-2	2300	350	703-30	N/A	N/S	122-61	N/R	N/S	N/R	125-50	0.028	N/R
R80321	2300	350	703-41	N/A	6-506	122-63	N/S	N/S	N/S	125-35	0.031	N/S
R80328	4175	650	703-40	N/A	6-511	122-62	N/S	N/S	N/S	125-50	0.04	RED
R80330	4150	850	703-35	N/A	6-504	122-88	122-94	N/S	N/S	125-65	0.04	PINK
R80340	4500	1050	37-1539	37-933	6-504	122-84	122-84	N/S	N/S	125-65 (15)	0.035	0.035
R80340-1	4500	1050	37-1539	37-933	6-504	122-84	122-84	N/S	N/S	125-65 (15)	0.035	0.035
R80340-2	4500	1050	37-1539	37-933	6-518-2	122-88	122-88	N/S	N/S	125-65 (15)	0.035	0.035
R80341	4160	390	3-720	N/A	6-506	122-54	34R9716-32	N/S	N/S	125-65	0.059	YELLOW
R80364	4150	450	703-28	N/A	6-511	122-59	N/S	N/S	N/S	125-85	0.021	Plain
R80364-1	4150	450	703-28	N/A	6-511	122-59	N/S	N/S	N/S	125-85		
R80378	4150	750	703-48	N/A	6-511	122-59	122-72	N/S	N/S	125-25 (30)	0.035	PINK
R80378-1	4150	750	703-48	N/A	6-511	122-56	122-73	N/S	N/S	125-25 (30)	0.031	PINK
R80382	2300	350	703-49	N/A	6-511	122-71	N/S	N/S	N/S	125-35	0.031	N/S
R80383	4160	650	703-47	N/A	6-511	122-68	N/S	N/S	N/S	125-25	0.035	BROWN
R80385	2300	350	703-41	N/A	6-506	122-75	N/S	N/S	N/S	125-25	0.028	N/S
R80386	2300	350	703-49	N/A	6-511	122-61	N/S	N/S	N/S	125-35	0.028	N/S
R80390	4175	650	703-50	N/A	6-511	122-61	34R9716-22	N/S	N/S	125-25	0.04	RED
R80391	4160	700	703-33	N/A	6-511	122-69	N/S	N/S	N/S	125-25	0.035	YELLOW
R80402	2300	450	703-36	N/A	6-511	122-75	N/S	N/S	N/S	125-25	0.028	N/S
R80402-1	2300	500	703-36	N/A	6-511	122-75	N/R	N/S	N/R	125-45	0.028	N/R
R80403	4160	600	703-29	N/A	6-511	122-64	N/S	N/S	N/S	125-25	0.025	RED
R80403-1	4160	600	703-29	N/A	6-511	122-64	N/S	N/S	N/S	125-25	0.032	RED
R80408	4150	715	703-45	N/A	6-504	122-73	-1	N/S	N/S	125-25	0.031	YELLOW
R80427	4150	700	3-485	N/A	6-504	122-74	122-84	N/S	N/S	125-45	0.037	0.037
R80431	4160	550	37-119	37-933	6-506	122-60	134-9	N/S	N/S	125-65	0.025	Plain

Primary Bowl Gasket	Primary Metering Block Gasket	Secondary Bowl Gasket	Secondary Metering Block Gasket	Secondary Metering Plate Gasket	Primary Fuel Bowl	Secondary Fuel Bowl	Throttle Body & Shaft Assembly	Venturii Diameter Primary	Venturii Diameter Secondary	Throttle Bore Diameter Primary	Throttle Bore Diameter Secondary
108-92-2	108-91-2	108-90-2	108-90-2	108-27-2	N/S	N/S	N/S	1-13/32	1-13/64	1-3/8	2
N/A	N/A	N/A	N/A	N/A	N/S	N/S	N/S	1-3/8	1-3/8	1-11/16	1-11/16
108-56-2	108-55-2	108-90-2	108-90-2	108-13-2	N/S	N/S	N/S	1-1/4	1-5/16	1-9/16	1-9/16
108-56-2	108-55-2	108-90-2	108-90-2	108-13-2	N/S	N/S	N/S	1-1/4	1-5/16	1-9/16	1-9/16
108-56-2	108-55-2	108-90-2	108-90-2	108-13-2	N/S	N/S	N/S	1-1/4	1-5/16	1-9/16	1-9/16
108-56-2	108-55-2	108-90-2	108-90-2	108-13-2	N/S	N/S	N/S	1-1/4	1-5/16	1-9/16	1-9/16
108-92-2	108-35-2	108-90-2	108-90-2	108-27-2	N/S	N/S	N/S	1-13/32	1-13/64	1-3/8	2
N/A	N/A	N/A	N/A	N/A	N/S	N/S	N/S	1-9/16	1-9/16	1-3/4	1-3/4
108-83-2	108-89-2	108-83-2	108-89-2	N/R	134-108	134-112	N/S	1.69	1.69	2	2
108-83-2	108-91-2	108-83-2	108-89-2	N/R	134-108	134-112	N/S	1-11/16	1-11/16	2	2
108-83-2	108-91-2	108-83-2	108-89-2	N/R	134-108	134-112	N/S	1-11/16	1-11/16	2	2
N/A	N/A	N/A	N/A	108-13-2	N/S	N/S	N/S	1-1/4	1-5/16	1-9/16	1-9/16
N/A	N/A	N/A	N/A	N/A	N/S	N/S	N/S	1-3/16	N/R	1-1/2	N/R
N/A	N/A	N/A	N/A	108-13-2	N/S	N/S	N/S	1-3/32	1-3/32	1-1/2	1-1/2
N/A	N/A	N/A	N/A	108-13-2	N/S	N/S	N/S	1-1/4	1-5/16	1-9/16	1-9/16
N/A	N/A	N/A	N/A	N/A	N/S	N/S	N/S	1-3/8	1-3/8	1-11/16	1-11/16
N/A	N/A	N/A	N/A	108-27-2	N/S	N/S	N/S	1-13/64	1-13/64	1-3/8	2
N/A	N/A	N/A	N/A	N/A	N/S	N/S	N/S	1-9/16	1-9/16	1-3/4	1-3/4
N/A	N/A	N/A	N/A	N/A	N/S	N/S	N/S	1-3/16	N/R	1-1/2	N/R
N/A	N/A	N/A	N/A	N/A	N/S	N/S	N/S	1-3/16	N/R	1-1/2	N/R
N/A	N/A	N/A	N/A	108-27-2	N/S	N/S	N/S	1-1/4	1-5/16	1-9/16	1-9/16
N/A	N/A	N/A	N/A	N/A	N/S	N/S	N/S	1-3/8	N/R	1-11/16	N/R
N/A	N/A	N/A	N/A	108-13-2	N/S	N/S	N/S	1-1/4	1-5/16	1-9/16	1-9/16
N/A	N/A	N/A	N/A	108-13-2	N/S	N/S	N/S	1-1/4	1-5/16	1-9/16	1-9/16
N/A	N/A	N/A	N/A	108-13-2	N/S	N/S	N/S	1-1/4	1-5/16	1-9/16	1-9/16
N/A	N/A	N/A	N/A	108-13-2	N/S	N/S	N/S	1-1/4	1-5/16	1-9/16	1-9/16
N/A	N/A	N/A	N/A	108-13-2	N/S	N/S	N/S	1-1/4	1-5/16	1-9/16	1-9/16
N/A	N/A	N/R	N/R	N/R	N/S	N/R	N/S	1-3/16	N/R	1-1/2	N/R
N/A	N/A	N/R	N/R	N/R	N/S	N/R	N/S	1-3/16	N/R	1-1/2	N/R
N/A	N/A	N/A	N/A	N/A	N/S	N/S	N/S	1-3/16	N/R	1-1/2	N/R
N/A	N/A	N/A	N/A	108-27-2	N/S	N/S	N/S	1-13/64	1-13/32	1-3/8	2
N/A	N/A	N/A	N/A	N/A	N/S	N/S	N/S	1-9/16	1-9/16	1-3/4	1-3/4
N/A	N/A	N/A	N/A	N/A	134-108	134-112	N/S	1-11/16	1-11/16	2	2
N/A	N/A	N/A	N/A	N/R	134-108	134-112	N/R	1-11/16	1-11/16	2	2
N/A	N/A	N/A	N/A	N/R	134-108	134-112	N/R	1-11/16	1-11/16	2	2
N/A	N/A	N/A	N/A	108-27-2	N/S	N/S	N/S	1-1/16	1-1/16	1-7/16	1-7/16
N/A	N/A	N/A	N/A	108-27-2	N/S	N/S	N/S	1-3/32	1-3/32	1-1/2	1-1/2
N/A	N/A	N/A	N/A	N/A	N/S	N/S	N/S	1-3/8	1-7/16	1-11/16	1-11/16
N/A	N/A	N/A	N/A	N/A	N/S	N/S	N/S	1-3/8	1-7/16	1-11/16	1-11/16
N/A	N/A	N/A	N/A	N/A	N/S	N/S	N/S	1-3/16	N/R	1-1/2	N/R
N/A	N/A	N/A	N/A	108-13-2	N/S	N/S	N/S	1-1/4	1-5/16	1-9/16	1-9/16
N/A	N/A	N/A	N/A	N/A	N/S	N/S	N/S	1-3/16	N/R	1-1/2	N/R
N/A	N/A	N/A	N/A	N/A	N/S	N/S	N/S	1-3/16	N/R	1-1/2	N/R
N/A	N/A	N/A	N/A	108-27-2	N/S	N/S	N/S	1-13/64	1-13/32	1-3/8	2
N/A	N/A	N/A	N/A	108-27-2	N/S	N/S	N/S	1-9/16		1-3/4	
N/A	N/A	N/A	N/A	N/A	N/S	N/S	N/S	1-9/16	N/R	1-3/4	N/R
N/A	N/A	N/R	N/R	N/R	N/S	N/R	N/S	1-9/16	N/R	1-3/4	N/R
N/A	N/A	N/A	N/A	108-27-2	N/S	N/S	N/S	1-1/4	1-5/16	1-9/16	1-9/16
N/A	N/A	N/A	N/A	108-27-2	N/S	N/S	N/S	1-1/4	1-5/16	1-9/16	1-9/16
N/A	N/A	N/A	N/A	N/A	N/S	N/S	N/S	1-5/16	1-3/8	1-11/16	1-11/16
N/A	N/A	N/A	N/A	N/A	N/S	N/S	N/S	1-5/16	1-3/8	1-11/16	1-11/16
108-83-2	108-89-2	108-90-2	108-90-2	108-27-2	N/S	N/S	N/S	1-3/16	1-1/4	1-1/2	1-1/2

Carb Number	Carb Model Number	CFM	Renew Kit	Trick Kit	Needle & Seat	Primary Main Jet	Secondary Main Jet or plate	Primary Metering Block	Secondary Metering Block	Power Valve	Pump Nozzle Size, Primary	Secondary Nozzle Size or Spring Color
R80432	4160	550	37-119	37-933	6-506	122-60	134-9	N/S	N/S	125-65	0.025	Plain
R80434	4160	750	703-55	N/A	N/A	122-69	N/S	N/S	N/S	125-45	0.035	ORANGE/RED
R80436	4150	850	37-1539	37-933	6-504	122-80	122-80	N/S	N/A	125-65 (22)	0.04	Pink
R80443	4150	850	703-58	N/A	6-504	122-88	122-96	N/S	N/S	125-65 (15)	0.031	0.031
R80444	4150	850	703-35	N/A	6-504	122-88	122-94	N/S	N/S	125-65 (22)	0.04	YELLOW
R80450	4160	600	37-1536	37-933	6-506	N/A	134-39	N/S	N/S	125-208	0.031	Black
R80451	4160	600	37-1536	37-933	6-506	N/A	134-39	N/S	N/S	125-208	0.031	Black
R80452	4160	600	37-1536	37-933	6-506	N/A	134-39	N/S	N/S	125-208	0.031	Black
R80453	4160	600	37-1536	37-933	6-506	N/A	134-39	N/S	N/S	125-208	0.031	Black
R80454	4160	600	37-1536	37-933	6-506	N/A	134-39	N/S	N/S	125-208	0.031	Black
R80456-1	4160	600	703-47	N/A	6-511	122-65	N/S	N/S	N/S	125-25	0.025	BROWN
R80457	4160	600	37-119	37-933	6-506	122-69	134-39	134-128	134-39	125-65	0.031	Black
R80457-1	4160	600	37-119	37-933	6-506	122-64	134-39	134-128	134-39	125-65	0.031	Black
R80457-2	4160	600	37-119	37-933	6-506	122-64	134-39	134-128S	134-39	125-65	0.031	Black
R80457-3	4160	600	37-119	37-933	6-506	122-64	134-39	134-128S	134-39	125-65	0.031	Black
R80457-4	4160	600	37-119	37-933	6-506	122-64	134-39	134-128S	134-39	125-65	0.031	Black
R80457-5	4160	600	37-119	37-933	6-511	122-64	134-39	134-128S	134-39	125-65	0.031	Black
R80457-6	4160	600	37-119	37-933	6-506	122-64	134-39	134-128S	134-39	125-65	0.031	Black
R80457-7	4160	600	37-119	37-933	6-511	122-64	134-39	134-128S	134-39	125-65	0.031	Black
R80460	4160	600	37-1536	37-933	6-506	N/A	134-39	N/S	N/S	125-208	0.031	Black
R80466	4150	800	3-485	N/A	6-504	122-72	122-87	N/S	N/S	125-45	0.031	0.031
R80473	4160	600	703-29	N/A	6-511	122-64	N/S	N/S	N/S	125-25	0.025	RED
R80487	4160	600	703-66	N/A	6-511	122-68	N/S	N/S	N/S	125-45	0.037	RED
R80491	4175	650	37-1537	37-933	6-511	N/A	134-21	N/S	134-21	N/S	0.037	Black
R80492	4160	600	703-29	N/A	6-511	122-68	N/S	N/S	N/S	N/A	0.037	Red
R80492-1	4160	600	703-29	N/A	6-511	122-68	N/S	N/S	N/S	N/A	0.037	
R80496	4150	950	37-1539	37-933	6-518-2	122-79	122-79	N/A	N/A	125-165 (both)	0.031	0.031
R80496-1	4150	950	37-1539	37-933	6-518-2	122-79	122-79	134-69	134-69	125-165 (both)	0.031	0.031
R80497	4150	950	37-1539	37-933	6-518-2	122-78	122-78	N/A	N/A	125-165 (both)	0.031	Brown
R80497-1	4150	950	37-1539	37-933	6-518-2	122-78	122-78	N/A	N/A	125-165 (both)	0.031	Brown
R80498	4150	950	37-1539	37-933	6-519-2	122-144	122-144	N/A	N/A	125-155 (both)	0.055	0.055
R80498-1	4150	950	37-1539	37-933	6-519-2	122-144	122-144	N/A	N/A	125-155 (both)	0.055	0.055
R80502	2300	500	3-474	N/A	6-504	122-71	N/S	N/S	N/S	125-35	0.047	N/S
R80507	4150	390	37-1539	37-933	6-504	122-65	122-65	N/A	N/A	125-35 (22)	0.025	0.025
R80507-1	4150	390	37-1539	37-933	6-504	122-65	122-65	N/A	N/A	125-35 (22)	0.025	0.025
R80507-2	4150	390	37-1539	37-933	6-504	122-65	122-65	N/A	N/A	125-35 (22)	0.025	0.025
R80507-3	4150	390	37-1539	37-933	6-504	122-65	122-65	N/A	N/A	125-35 (22)		
R80508	4160	750	37-754	37-933	6-504	122-72	134-21	134-131	134-21	125-65	0.025	Plain
R80508-1	4160	750	37-754	37-933	6-504	122-72	134-21	134-131S	134-21	125-65	0.025	Plain
R80508-2	4160	750	37-754	37-933	6-504	122-72	134-21	134-131S	134-21	125-65	0.031	
R80508-3	4160	750	37-754	37-933	6-504	122-70	134-21	N/A	134-21	125-65	0.031	
R80508-4	4160	750	37-754	37-933	6-504	122-70	134-21	N/A	134-21	125-65	0.031	
R80508-5	4160	750	37-754	37-933	6-504	122-70	134-21	N/A	134-21	125-65	0.031	
R80508-6	4160	750	37-754	37-933	6-504	122-70	134-21	N/A	134-21	125-65	0.031	
R80508-7	4160	750	37-754	37-933	6-504	122-70	134-21	N/A	134-21	125-65	0.031	
R80509	4150	830	37-1539	37-933	6-504	122-86	122-86	N/A	N/A	125-65 (15)	0.028	0.029
R80509-1	4150	830	37-1539	37-933	6-504	122-86	122-86	N/A	N/A	125-65 (15)	0.028	0.029
R80509-2	4150	830	37-1539	37-933	6-504	122-86	122-86	N/A	N/A	125-65 (15)	0.028	0.029

Primary Bowl Gasket	Primary Metering Block Gasket	Secondary Bowl Gasket	Secondary Metering Block Gasket	Secondary Metering Plate Gasket	Primary Fuel Bowl	Secondary Fuel Bowl	Throttle Body & Shaft Assembly	Venturii Diameter Primary	Venturii Diameter Secondary	Throttle Bore Diameter Primary	Throttle Bore Diameter Secondary
108-83-2	108-89-2	108-90-2	108-90-2	108-27-2	N/S	N/S	N/S	1-3/16	1-1/4	1-1/2	1-1/2
N/A	N/A	N/A	N/A	108-13-2	N/S	N/S	N/S	1-3/8	1-3/8	1-11/16	1-11/16
108-83-2	108-89-2	108-83-2	108-89-2	N/R	134-103	134-102	N/A	1-9/16	1-9/16	1-3/4	1-3/4
N/A	N/A	N/A	N/A	N/R	N/S	N/S	N/S	1-9/16	1-9/16	1-3/4	1-3/4
N/A	N/A	N/A	N/A	N/A	N/S	N/S	N/S	1-9/16	1-9/16	1-3/4	1-3/4
108-83-2	108-89-2	108-90-2	108-90-2	N/R	N/S	N/S	N/S	1-1/4	1-5/16	1-9/16	1-9/16
108-83-2	108-89-2	108-90-2	108-90-2	N/R	N/S	N/S	N/S	1-1/4	1-5/16	1-9/16	1-9/16
108-83-2	108-89-2	108-90-2	108-90-2	N/R	N/S	N/S	N/S	1-1/4	1-5/16	1-9/16	1-9/16
108-83-2	108-89-2	108-90-2	108-90-2	N/R	N/S	N/S	N/S	1-1/4	1-5/16	1-9/16	1-9/16
108-83-2	108-89-2	108-90-2	108-90-2	N/R	N/S	N/S	N/S	1-1/4	1-5/16	1-9/16	1-9/16
N/A	N/A	N/A	N/A	108-13-2	N/S	N/S	N/S	1-1/4	1-5/16	1-9/16	1-9/16
108-83-2	108-89-2	108-90-2	108-90-2	N/R	134-101	134-105	N/S	1-1/4	1-5/16	1-9/16	1-9/16
108-83-2	108-89-2	108-90-2	108-90-2	N/R	134-101	134-105	N/A	1-1/4	1-5/16	1-9/16	1-9/16
108-83-2	108-89-2	108-90-2	108-90-2	N/R	134-101S	134-105S	N/A	1-1/4	1-5/16	1-9/16	1-9/16
108-83-2	108-89-2	108-90-2	108-90-2	N/R	134-101S	134-105S	N/A	1-1/4	1-5/16	1-9/16	1-9/16
108-83-2	108-89-2	108-90-2	108-90-2	N/R	134-101S	134-105S	N/A	1-1/4	1-5/16	1-9/16	1-9/16
108-83-2	108-89-2	108-90-2	108-90-2	N/R	134-101S	134-105S	N/A	1-1/4	1-5/16	1-9/16	1-9/16
108-83-2	108-89-2	108-90-2	108-90-2	N/R	134-101S	134-105S	N/A	1-1/4	1-5/16	1-9/16	1-9/16
108-83-2	108-89-2	108-90-2	108-90-2	N/R	134-281S	134-282S	112-113	1-1/4	1-5/16	1-9/16	1-9/16
108-83-2	108-89-2	108-90-2	108-90-2	N/R	N/S	N/S	N/S	1-1/4	1-5/16	1-9/16	1-9/16
N/A	N/A	N/A	N/A	N/A	N/S	N/S	N/S	1-3/8	1-7/16	1-11/16	1-11/16
N/A	N/A	N/A	N/A	108-27-2	N/S	N/S	N/S	1-1/4	1-5/16	1-9/16	1-9/16
N/A	N/A	N/A	N/A	108-27-2	N/S	N/S	N/S	1-1/4	1-5/16	1-9/16	1-9/16
108-92-2	108-91-2	108-90-2	108-90-2	108-27-2	N/S	N/A	N/S	1-13/32	1-13/64	1-3/8	2
N/A	N/A	N/A	N/A	108-27-2	N/S	N/S	N/S	1-1/4	1-5/16	1-9/16	1-9/16
N/A	N/A	N/A	N/A	108-27-2	N/S	N/S	N/S	1-1/4	1-5/16	1-9/16	1-9/16
108-83-2	108-89-2	108-83-2	108-89-2	N/R	134-108	134-112	N/A	1-3/8	1-3/8	1-3/4	1-3/4
108-83-2	108-89-2	108-83-2	108-89-2	N/R	134-108	134-112	112-122	1-3/8	1-3/8	1-3/4	1-3/4
108-83-2	108-89-2	108-83-2	108-89-2	N/R	134-108	34R11442	N/A	1-3/8	1-3/8	1-3/4	1-3/4
108-83-2	108-89-2	108-83-2	108-89-2	N/R	134-108	N/A	N/A	1-3/8	1-3/8	1-3/4	1-3/4
108-83-2	108-89-2	108-83-2	108-89-2	N/R	N/A	N/A	N/A	1-3/8	1-3/8	1-3/4	1-3/4
108-83-2	108-89-2	108-83-2	108-89-2	N/R	N/A	N/A	N/A	1-3/8	1-3/8	1-3/4	1-3/4
N/A	N/A	N/A	N/A	N/A	N/S	N/S	N/S	1-3/8	N/R	1-11/16	N/R
108-83-2	108-89-2	108-83-2	108-89-2	N/R	134-103	134-104	N/A	1-1/16	1-1/16	1-7/16	1-7/16
108-83-2	108-89-2	108-83-2	108-89-2	N/R	134-103	134-104	N/A	1-1/16	1-1/16	1-7/16	1-7/16
108-83-2	108-89-2	108-83-2	108-89-2	N/R	134-103	134-104	N/A	1-1/16	1-1/16	1-7/16	1-7/16
108-83-2	108-89-2	108-90-2	108-90-2	108-27-2	134-103	134-102	N/S	1-3/8	1-7/16	1-11/16	1-11/16
108-83-2	108-89-2	108-90-2	108-90-2	108-27-2	134-103S	134-102S	N/A	1-3/8	1-7/16	1-11/16	1-11/16
108-83-2	108-89-2	108-90-2	108-90-2	108-27-2	134-103S	134-102S	N/A	1-3/8	1-7/16	1-11/16	1-11/16
108-83-2	108-89-2	108-90-2	108-90-2	108-27-2	134-103S	134-102S	N/A	1-3/8	1-7/16	1-11/16	1-11/16
108-83-2	108-89-2	108-90-2	108-90-2	108-27-2	134-103S	134-102S	N/A	1-3/8	1-7/16	1-11/16	1-11/16
108-83-2	108-89-2	108-90-2	108-90-2	108-27-2	134-103S	134-102S	N/A	1-3/8	1-7/16	1-11/16	1-11/16
108-83-2	108-89-2	108-90-2	108-90-2	108-27-2	134-103S	134-102S	N/A	1-3/8	1-7/16	1-11/16	1-11/16
108-83-2	108-89-2	108-90-2	108-90-2	108-27-2	134-103S	134-102S	112-117	1-3/8	1-7/16	1-11/16	1-11/16
108-83-2	108-89-2	108-83-2	108-89-2	N/R	134-103	134-104	N/S	1-9/16	1-9/16	1-11/16	1-11/16
108-83-2	108-89-2	108-83-2	108-89-2	N/R	134-103	134-104	N/S	1-9/16	1-9/16	1-11/16	1-11/16
108-83-2	108-89-2	108-83-2	108-89-2	N/R	134-103	134-104	N/S	1-9/16	1-9/16	1-11/16	1-11/16

Carb Number	Carb Model Number	CFM	Renew Kit	Trick Kit	Needle & Seat	Primary Main Jet	Secondary Main Jet or plate	Primary Metering Block	Secondary Metering Block	Power Valve	Pump Nozzle Size, Primary	Secondary Nozzle Size or Spring Color
R80511	4150	830	37-1539	37-933	6-518-2	122-84	122-84	N/A	N/A	125-65 (15)	0.028	0.029
R80511-1	4150	830	37-1539	37-933	6-518-2	122-84	122-84	N/A	N/A	125-65 (15)	0.028	0.029
R80511-2	4150	830	37-1539	37-933	6-518-2	122-84	122-84	N/A	N/A	125-65 (15)	0.028	0.029
R80512	4150	1000	37-1539	37-933	6-518-2	122-84	122-84	N/A	N/A	125-65 (15)	0.031	Brown
R80513	4150	1000	37-1539	37-933	6-518-2	122-84	122-84	N/A	N/A	125-65 (15)	0.031	0.036
R80513-1	4150	1000	37-1539	37-933	6-518-2	122-84	122-84	N/A	N/A	125-65 (15)	0.031	0.036
R80514	4150	1000	37-1539	37-933	6-518-2	122-84	122-88	N/A	N/A	125-65 (15)	0.031	0.036
R80514-1	4150	1000	37-1539	37-933	6-518-2	122-84	122-88	N/A	N/A	125-65 (15)	0.031	0.036
R80519	4150	1000	37-1539	37-933	6-518-2	122-84	122-88	N/A	N/A	125-65 (15)	0.031	Brown
R80528	4150	750	37-1539	37-933	6-504	122-72	122-84	N/A	N/A	125-65	0.031	0.031
R80528-1	4150	750	37-1539	37-933	6-504	122-73	122-73	N/A	N/A	125-65 (15)	0.031	0.031
R80528-2	4150	750	37-1539	37-933	6-504	122-73	122-73	134-67	134-67	125-65 (15)	0.031	0.031
R80529	4150	750	37-1539	37-933	6-504	122-72	122-84	N/A	N/A	125-65	0.031	Brown
R80529-1	4150	750	37-1539	37-933	6-504	122-72	122-84	N/A	N/A	N/R	0.031	Brown
R80531	4150	850	37-1539	37-933	6-504	122-78	122-82	N/A	N/A	125-45 (22)	0.04	Pink
R80532	4500	1250	37-1539	37-933	6-518-2	122-101	122-101	N/A	N/A	122-55	0.035	0.035
R80532-1	4500	1250	37-1539	37-933	6-518-2	122-97	122-97	N/A	N/A	N/R	0.035	0.037
R80533	4500	1250	37-1539	37-933	6-518-2	122-97	122-97	N/A	N/A	N/R	0.035	0.037
R80533-1	4500	1250	37-1539	37-933	6-518-2	122-97	122-97	N/A	N/A	N/R	0.035	0.037
R80535	4150	750	37-1539	37-933	6-519-2	122-132	122-132	N/A	N/A	125-55	0.045	0.045
R80535-1	4150	750	37-1539	37-933	6-519-2	122-144	122-144	N/A	N/A	125-55	0.045	0.045
R80535-2	4150	750	37-1539	37-933	6-519-2	122-144	122-144	N/A	N/A	125-55	0.045	0.045
R80535-3	4150	750	37-1539	37-933	6-519-2	122-144	122-144	N/A	N/A	125-55	0.045	0.045
R80537	4150	750	3-485	N/A	6-504	122-73	122-81	N/S	N/A	125-65	0.028	0.031
R80540	4150	600	37-1539	37-933	6-518-2	122-70	122-70	N/A	N/A	125-65	0.028	0.029
R80540-1	4150	600	37-1539	37-933	6-518-2	122-70	122-70	N/A	N/A	125-65	0.028	0.029
R80540-2	4150	600	37-1539	37-933	6-518-2	122-70	122-70	N/A	N/A	125-65	0.028	0.029
R80541	4150	650	37-1539	37-933	6-518-2	122-70	122-70	N/A	N/A	125-65	0.028	0.029
R80541-1	4150	650	37-1539	37-933	6-518-2	122-70	122-70	N/A	N/A	125-65	0.028	0.029
R80541-2	4150	650	37-1539	37-933	6-518-2	122-70	122-70	134-68	134-68	125-65	0.028	0.029
R80542	4150	650	37-1539	37-933	6-519-2	122-90	122-90	N/A	N/A	125-65	0.055	0.055
R80542-1	4150	650	37-1539	37-933	6-519-2	122-90	122-90	N/A	N/A	125-65	0.055	0.055
R80551	4160	600	703-1	N/A	6-511	122-63	N/A	N/A	N/A	125-25	0.037	Red
R80551-1	4160	600	703-1	N/A	6-511	122-63	N/A	N/A	N/A	125-25	0.037	Red
R80552	4175	650	703-34	N/A	6-511	122-61	N/A	N/A	N/A	125-50	0.04	Red
R80555	4175	650	37-1537	37-933	6-510	122-62	N/A	N/A	N/A	125-65	0.04	Yellow
R80555-1	4175	650	37-1537	37-933	6-510	122-62	N/A	N/A	N/A	125-65	0.04	Yellow
R80556	4500	1150	37-1539	37-933	6-518-2	122-92	122-92	N/A	N/A	125-55	0.035	0.035
R80556-1	4500	1150	37-1539	37-933	6-518-2	122-90	122-90	N/A	N/A	125-55	0.035	0.035
R80557	4150	750	703-47	N/A								
R80558	4150	830	703-4	N/A		122-85	122-85					
R80559	4150	600	N/A	N/A	6-504	122-67	122-74	N/A	N/A	125-65	0.028	0.032
R80570	4160	570	37-934	N/A	6-506	122-54	122-65	N/S	N/S	125-85	0.031	Plain
R80572	4150	700	N/A	37-933	6-504	122-72	122-82	N/A	N/A	125-105	0.028	0.031
R80573	4150	750	N/A	37-933	6-504	122-74	122-84	N/A	N/A	125-105	0.028	0.031
R80574	4150	800	N/A	N/A	6-504	122-74	122-82	N/A	N/A	125-105	0.031	0.031
R80575	4150	600	37-1544	37-933	6-518-2	122-73	122-73	N/A	N/A	125-105	0.028	0.029
R80575-1	4150	600	37-1544	37-933	6-518-2	122-73	122-73	N/A	N/A	125-105	0.028	0.029
R80576	4150	750	N/A	37-933	6-504	122-76	122-87	N/A	N/A	125-105	0.031	0.031
R80576-1	4150	750	37-1544	37-933	6-504	122-76	122-87	N/A	N/A	125-105	0.031	0.031
R80577	4150	850	37-1544	37-933	6-518-2	122-80	122-86	N/A	N/A	125-105	0.031	0.031
R80577-1	4150	850	37-1544	37-933	6-504	122-80	122-86	N/A	N/A	125-105	0.031	0.031
R80578	4500	1150	N/A	37-933	6-518-2	122-99	122-99	N/A	N/A	N/R	0.035	0.035

Primary Bowl Gasket	Primary Metering Block Gasket	Secondary Bowl Gasket	Secondary Metering Block Gasket	Secondary Metering Plate Gasket	Primary Fuel Bowl	Secondary Fuel Bowl	Throttle Body & Shaft Assembly	Venturii Diameter Primary	Venturii Diameter Secondary	Throttle Bore Diameter Primary	Throttle Bore Diameter Secondary
108-83-2	108-89-2	108-83-2	108-89-2	N/R	134-108	134-112	N/S	1-9/16	1-9/16	1-11/16	1-11/16
108-83-2	108-89-2	108-83-2	108-89-2	N/R	134-108	134-112	N/S	1-9/16	1-9/16	1-11/16	1-11/16
108-83-2	108-89-2	108-83-2	108-89-2	N/R	134-108	134-112	N/S	1-9/16	1-9/16	1-11/16	1-11/16
108-83-2	108-89-2	108-83-2	108-89-2	N/R	134-108	N/A	N/A	1-9/16	1-9/16	1-3/4	1-3/4
108-83-2	108-89-2	108-83-2	108-89-2	N/R	134-108	134-112	N/A	1-9/16	1-9/16	1-3/4	1-3/4
108-83-2	108-89-2	108-83-2	108-89-2	N/R	134-108	134-112	N/A	1-9/16	1-9/16	1-3/4	1-3/4
108-83-2	108-89-2	108-83-2	108-89-2	N/R	134-108	134-112	N/A	1-9/16	1-9/16	1-3/4	1-3/4
108-83-2	108-89-2	108-83-2	108-89-2	N/R	134-108	134-112	N/A	1-9/16	1-9/16	1-3/4	1-3/4
108-83-2	108-89-2	108-83-2	108-89-2	N/R	134-108	N/A	N/A	1-9/16	1-9/16	1-3/4	1-3/4
108-83-2	108-89-2	108-83-2	108-89-2	N/R	134-108	134-112	N/A	1-3/8	1-3/8	1-11/16	1-11/16
108-83-2	108-89-2	108-83-2	108-89-2	N/R	134-108	134-112	N/A	1-3/8	1-3/8	1-11/16	1-11/16
108-83-2	108-89-2	108-83-2	108-89-2	N/R	134-103	134-112	112-120	1-3/8	1-3/8	1-11/16	1-11/16
108-83-2	108-89-2	108-83-2	108-89-2	N/R	134-108	N/A	N/A	1-3/8	1-3/8	1-11/16	1-11/16
108-83-2	108-89-2	108-83-2	108-89-2	N/R	134-108	N/A	N/A	1-3/8	1-3/8	1-11/16	1-11/16
108-83-2	108-89-2	108-83-2	108-89-2	N/R	N/A	N/A	N/A	1-3/8	1-3/8	1-11/16	1-11/16
108-83-2	108-36-2	108-36-2	108-36-2	N/R	N/A	N/A	N/S	1.88	1.88	2-1/8	2-1/8
108-120	108-121	108-120	108-121	N/R	134-108	134-112	N/S	1.88	1.88	2-1/8	2-1/8
108-120	108-121	108-120	108-121	N/R	134-108	134-112	N/S	1.88	1.88	2-1/8	2-1/8
108-120	108-121	108-120	108-121	N/R	134-108	134-112	N/S	1.88	1.88	2-1/8	2-1/8
108-83-2	108-89-2	108-83-2	108-89-2	N/R	134-108	134-112	N/A	1-3/8	1-3/8	1-11/16	1-11/16
108-83-2	108-89-2	108-83-2	108-89-2	N/R	134-108	134-112	N/A	1-3/8	1-3/8	1-11/16	1-11/16
108-83-2	108-89-2	108-83-2	108-89-2	N/R	134-108	134-112	N/A	1-3/8	1-3/8	1-11/16	1-11/16
108-83-2	108-89-2	108-83-2	108-89-2	N/R	134-108	134-112	N/A	1-3/8	1-3/8	1-11/16	1-11/16
N/A	N/A	N/A	N/A	N/R	134-108	134-112	N/S	1-3/8	1-3/8	1-11/16	1-11/16
108-83-2	108-89-2	108-83-2	108-89-2	N/R	134-108	134-112	N/A	1-1/4	1-1/4	1-9/16	1-9/16
108-83-2	108-89-2	108-83-2	108-89-2	N/R	134-108	134-112	N/A	1-1/4	1-1/4	1-9/16	1-9/16
108-83-2	108-89-2	108-83-2	108-89-2	N/R	134-108	134-112	N/A	1-1/4	1-1/4	1-9/16	1-9/16
108-83-2	108-89-2	108-83-2	108-89-2	N/R	134-108	134-112	N/A	1-1/4	1-1/4	1-11/16	1-11/16
108-83-2	108-89-2	108-83-2	108-89-2	N/R	134-108	134-112	N/A	1-1/4	1-1/4	1-11/16	1-11/16
108-83-2	108-89-2	108-83-2	108-89-2	N/R	134-108	134-112	112-121	1-1/4	1-1/4	1-11/16	1-11/16
108-83-2	108-89-2	108-83-2	108-89-2	N/R	N/A	N/A	N/A	1-1/4	1-1/4	1-11/16	1-11/16
108-83-2	108-89-2	108-83-2	108-89-2	N/R	N/A	N/A	N/A	1-1/4	1-1/4	1-11/16	1-11/16
N/A	N/A	N/A	N/A	N/A	N/A	N/A	N/A	1-3/8	1-7/16	1-9/16	1-9/16
N/A	N/A	N/A	N/A	N/A	N/A	N/A	N/A	1-3/8	1-7/16	1-9/16	1-9/16
N/A	N/A	N/A	N/A	N/A	N/A	N/A	N/A	1-13/64	1-13/32	1-3/8	2
108-92-2	108-91-2	108-90-2	108-90-2	108-27-2	N/A	N/A	N/A	1-13/64	1-13/32	1-3/8	2
108-92-2	108-91-2	108-90-2	108-90-2	108-27-2	N/A	N/A	N/A	1-13/64	1-13/32	1-3/8	2
108-83-2	108-36-2	108-36-2	108-36-2	N/R	N/A	N/A	N/S	1.83	1.83	2	2
108-120	108-121	108-120	108-121	N/R	N/A	N/A	N/S	2	2	1.83	1.83
								1-3/8	1-3/8	1-11/16	1-11/16
								1-9/16	1-9/16	1-11/16	1-11/16
N/A	N/A	N/A	N/A	N/R	34R11341-1	34R11335	12R11335A	1-1/4	1-5/16	1-9/16	1-9/16
108-83-2	108-89-2	108-83-2	108-89-2	N/R	N/S	N/S	N/S	1-3/32	1-3/32	1-1/2	1-1/2
108-83-2	108-89-2	108-83-2	108-89-2	N/R	134-103S	134-104S	N/A	1-5/16	1-3/8	1-11/16	1-11/16
108-83-2	108-89-2	108-83-2	108-89-2	N/R	134-103S	134-104S	N/A	1-3/8	1-3/8	1-11/16	1-11/16
108-83-2	108-89-2	108-83-2	108-89-2	N/R	134-103S	134-104S	N/A	1-3/8	1-7/16	1-11/16	1-11/16
108-83-2	108-89-2	108-83-2	108-89-2	N/R	N/A	N/A	N/A	1-1/4	1-1/4	1-9/16	1-9/16
108-83-2	108-89-2	108-83-2	108-89-2	N/R	134-102S	134-104S	N/A	1-1/4	1-1/4	1-9/16	1-9/16
108-83-2	108-89-2	108-83-2	108-89-2	N/R	134-108	134-112	N/A	1-3/8	1-3/8	1-11/16	1-11/16
108-83-2	108-89-2	108-83-2	108-89-2	N/R	134-108	134-112	N/A	1-3/8	1-3/8	1-11/16	1-11/16
108-83-2	108-89-2	108-83-2	108-89-2	N/R	134-108	134-112	N/A	1 3/8	1 3/8	1 3/4	1 3/4
108-83-2	108-89-2	108-83-2	108-89-2	N/R	134-108	134-112	N/A	1 3/8	1 3/8	1 3/4	1 3/4
108-120	108-121	108-120	108-121	N/R	134-108	134-112	N/R	1-13/16	1-13/16	2	2

Carb Number	Carb Model Number	CFM	Renew Kit	Trick Kit	Needle & Seat	Primary Main Jet	Secondary Main Jet or plate	Primary Metering Block	Secondary Metering Block	Power Valve	Pump Nozzle Size, Primary	Secondary Nozzle Size or Spring Color
R80583-1	2300	500	37-1543	N/A	6-504	122-73	N/A	N/S	N/A	125-35	0.028	N/A
R80670	4160	670	37-935	37-933	6-506	122-65	122-68	N/S	N/S	125-65	0.031	Plain
R80670-1	4160	670	37-935	37-933	6-504	122-65	122-68	N/S	N/S	125-65	0.031	Plain
R80670-2	4160	670	37-935	37-933	6-504	122-65	122-68	134-57	134-58	125-65	0.031	Plain
R80672	4500	1050	37-1539	37-933	6-518-2	122-88	122-88	N/A	N/A	125-55 (both)	0.035	0.035
R80673	4500	1150	37-1539	37-933	6-518-2	122-90	122-90	N/A	N/A	125-55 (both)	0.035	0.035
R80674	4150	650	37-1539	37-933	6-518-2	122-70	122-70	N/A	N/A	125-65 (both)	0.028	0.029
R80675	4150	750	37-1539	37-933	6-504	122-73	122-73	N/A	N/A	125-65 (both)	0.031	0.031
R80676	4150	950	37-1539	37-933	6-518-2	122-78	122-78	N/A	N/A	125-65 (both)	0.031	0.031
R80681	4150	670	37-936	N/A	6-513	122-68	122-89	N/A	N/A	125-25	0.028	Black
R80770	4160	770	37-935	N/A	6-506	122-72	122-75	N/S	N/S	125-65	0.025	Plain
R80770-1	4160	770	37-935	37-933	6-504	122-72	122-75	134-59	134-60	125-65	0.025	
R80776	4150	600	37-485	37-933	6-504	122-66	122-73	N/A	N/A	125-65	0.028	0.031
R80777	4150	650	37-485	37-933	6-504	122-67	122-73	134-150	N/A	125-65	0.028	0.028
R80778	4150	700	37-485	37-933	6-504	122-69	122-78	N/A	N/A	125-65	0.028	0.031
R80779	4150	750	37-485	37-933	6-504	122-70	122-80	N/A	N/A	125-65	0.028	0.031
R80780	4150	800	37-485	37-933	6-504	122-71	122-85	N/A	N/A	125-65	0.031	0.031
R80781	4150	850	37-485	37-933	6-504	122-80	122-78	N/A	N/A	125-65 (15)	0.031	0.031
R80783	4150	650	N/A	N/A	6-504	122-67	122-73	34-150	N/A	125-65	0.028	Yellow
R80783-1	4150	650	N/A	N/A	6-504	122-67	122-73	34-150	N/A	125-65	0.028	Yellow
R80787-1	2300	350	N/A	N/A	6-504	122-77	N/A	134-276	N/A	125-45	0.021	N/R
R80801	4150	600	37-1548	N/A	6-518-2	122-68	122-68	N/S	N/S	122-65	0.032	0.032
R80802	4150	650	37-1548	N/A	6-518-2	122-70	122-70	N/S	N/S	122-65	0.032	0.032
R80803	4150	750	37-1548	N/A	6-518-2	122-76	122-76	N/S	N/S	122-65	0.032	0.032
R80804	4150	850	37-1548	N/A	6-518-2	122-84	122-84	N/S	N/S	122-45	0.032	0.032
R80805	4150	950	37-1548	N/A	6-518-2	122-92	122-92	N/S	N/S	122-45	0.032	0.032
R80870	4160	870	37-934	N/A	6-506	122-78	122-82	N/S	N/S	125-45	0.042	Plain
R80870-1	4160	870	37-934	37-933	6-504	122-78	122-82	N/S	N/S	125-45	0.042	
R81570	4150	570	37-934	37-933	6-506	122-54	122-65	N/A	N/A	125-85	0.031	Plain
R81670	4150	670	37-935	37-933	6-506	122-65	122-68	N/A	N/A	125-65	0.031	Plain
R81770	4150	770	37-935	37-933	6-506	122-72	122-75	N/A	N/A	125-65	0.025	Plain
R81870	4150	870	37-934	37-933	6-504	122-78	122-82	N/A	N/A	125-45 (22)	0.04	Plain
R81850	4160	600	37-119	37-933	6-506	122-66	134-9	134-128	134-9	125-65	0.025	Plain
R82010	2010	350	37-1541	N/A	6-504	122-58	N/A	N/R	N/R	125-65	0.035	N/R
R82011	2010	500	37-1541	N/A	6-504	122-80	N/A	N/R	N/R	125-65	0.035	N/R
R82012	2010	560	37-1541	N/A	6-504	122-80	N/A	N/R	N/R	125-65	0.035	N/R
R82020	2010	350	703-51	N/A	6-504	122-60	N/S	N/S	N/S	125-65	0.035	N/S
R82021	2010	500	703-51	N/A	6-504	122-80	N/S	N/S	N/S	125-65	0.035	N/S
R82028	2010	500	703-51	N/A	6-504	122-80	N/S	N/S	N/S	125-65	0.035	N/S
R82029	2010	500	703-51	N/A	6-504	122-81	N/S	N/S	N/S	125-65	0.035	N/S
R82750	4150	750	37-1539	37-933	6-504	122-75	122-76	N/A	N/A	125-45	0.031	Plain
R82751	4150	750	37-1539	37-933	6-504	122-75	122-80	N/A	N/A	125-45	0.031	0.028
R83310	4160	750	37-754	37-933	6-504	122-72	134-21	134-131	134-21	125-65	0.025	Plain
R83310-1	4160	750	37-754	37-933	6-504	122-72	134-21	134-131	134-21	125-65	0.025	Plain
R83311	4160	750	37-754	37-933	6-504	122-72	134-21	134-131	134-21	125-65	0.025	Plain
R83312	4160	750	37-754	37-933	6-504	122-72	134-21	134-131	134-21	125-65	0.025	Plain
R84010	4010	600	37-1541	N/A	6-504	122-67	122-75	N/R	N/R	125-65	0.026	Purple
R84010-1	4010	600	37-1541	N/A	6-504	122-67	122-75	N/R	N/R	125-65	0.035	Purple
R84010-2	4010	600	37-1541	N/A	6-504	122-67	122-75	N/R	N/R	125-65	0.035	Purple
R84010-3	4010	600	37-1541	N/A	6-504	122-63	122-75	N/R	N/R	125-65	0.035	Purple
R84011	4010	750	37-1541	N/A	6-504	122-75	122-75	N/R	N/R	125-65 (15)	0.026	Purple
R84011-1	4010	750	37-1541	N/A	6-504	122-75	122-75	N/R	N/R	125-65 (15)	0.035	Purple
R84011-2	4010	750	37-1541	N/A	6-504	122-75	122-75	N/R	N/R	125-65	0.035	Purple

Primary Bowl Gasket	Primary Metering Block Gasket	Secondary Bowl Gasket	Secondary Metering Block Gasket	Secondary Metering Plate Gasket	Primary Fuel Bowl	Secondary Fuel Bowl	Throttle Body & Shaft Assembly	Venturii Diameter Primary	Venturii Diameter Secondary	Throttle Bore Diameter Primary	Throttle Bore Diameter Secondary
108-83-2	108-89-2	N/A	N/A	N/A	N/S	N/A	N/S	1-3/8	N/R	1-11/16	N/R
108-83-2	108-89-2	108-83-2	108-89-2	N/R	N/S	N/S	N/S	1-1/4	1-5/16	1-9/16	1-9/16
108-83-2	108-89-2	108-83-2	108-89-2	N/R	N/S	N/S	N/S	1-1/4	1-5/16	1-9/16	1-9/16
108-83-2	108-89-2	108-83-2	108-89-2	N/R	N/S	N/S	112-114	1-1/4	1-5/16	1-9/16	1-9/16
108-120	108-121	108-120	108-121	N/R	N/A	N/A	N/S	1-11/16	1-11/16	2	2
108-120	108-121	108-120	108-121	N/R	N/A	N/A	N/S	2	2	1.83	1.83
108-83-2	108-89-2	108-83-2	108-89-2	N/R	N/A	N/A	N/A	1-1/4	1-1/4	1-11/16	1-11/16
108-83-2	108-89-2	108-83-2	108-89-2	N/R	N/A	N/A	N/A	1-3/8	1-3/8	1-11/16	1-11/16
108-83-2	108-89-2	108-83-2	108-89-2	N/R	N/A	N/A	N/A	1-3/8	1-3/8	1-3/4	1-3/4
108-83-2	108-89-2	108-83-2	108-89-2	N/R	N/A	N/A	N/A N/A	1-1/4	1-5/16	1-9/16	1-9/16
108-83-2	108-89-2	108-83-2	108-89-2	N/R	N/S	N/S	N/S	1-3/8	1-7/16	1-11/16	1-11/16
108-83-2	108-89-2	108-83-2	108-89-2	N/R	N/S	N/S	112-115	1-3/8	1-7/16	1-11/16	1-11/16
108-83-2	108-89-2	108-83-2	108-89-2	N/R	134-103	134-104	N/A	1-1/4	1-5/16	1-9/16	1-9/16
108-83-2	108-89-2	108-83-2	108-89-2	N/R	134-103	134-104	112-17	1-1/4	1-5/16	1-11/16	1-11/16
108-83-2	108-89-2	108-83-2	108-89-2	N/R	134-103	134-104	N/A	1-5/16	1-3/8	1-11/16	1-11/16
108-83-2	108-89-2	108-83-2	108-89-2	N/R	134-103	134-104	N/A	1-3/8	1-3/8	1-11/16	1-11/16
108-83-2	108-89-2	108-83-2	108-89-2	N/R	134-103	134-104	N/A	1-3/8	1-7/16	1-11/16	1-11/16
108-83-2	108-89-2	108-83-2	108-89-2	N/R	134-103	134-104	N/A	1-9/16	1-9/16	1-3/4	1-3/4
108-83-2	108-89-2	108-83-2	108-89-2	N/R	134-103S	134-104S		1-1/4	1-5/16	1-11/16	1-11/16
108-83-2	108-89-2	108-83-2	108-89-2	N/R	134-103S	134-104S		1-1/4	1-5/16	1-11/16	1-11/16
108-83-2	108-89-2	N/A	N/A	N/A	N/S	N/A	N/S	1-3/16	N/R	1-1/2	N/R
108-83-2	108-89-2	108-83-2	108-89-2	N/R	N/S	N/S	N/S	1-1/4	1-1/4	1-9/16	1-9/16
108-83-2	108-89-2	108-83-2	108-89-2	N/R	N/S	N/S	N/S	1-1/4	1-1/4	1-11/16	1-11/16
108-83-2	108-89-2	108-83-2	108-89-2	N/R	N/S	N/S	N/S	1-3/8	1-3/8	1-11/16	1-11/16
108-83-2	108-89-2	108-83-2	108-89-2	N/R	N/S	N/S	N/S	1-9/16	1-9/16	1-3/4	1-3/4
108-83-2	108-89-2	108-83-2	108-89-2	N/R	N/S	N/S	N/S	1.6	1.6	1-3/4	1-3/4
108-83-2	108-89-2	108-83-2	108-89-2	N/R	N/S	N/S	N/S	1-3/8	1-3/8	1-11/16	1-11/16
108-83-2	108-89-2	108-83-2	108-89-2	N/R	N/S	N/S	N/S	1-3/8	1-3/8	1-11/16	1-11/16
108-83-2	108-89-2	108-83-2	108-89-2	N/R	N/A	N/A	N/A	1-1/4	1-5/16	1-9/16	1-9/16
108-83-2	108-89-2	108-83-2	108-89-2	N/R	N/A	N/A	N/A	1-3/8	1-7/16	1-11/16	1-11/16
108-83-2	108-89-2	108-83-2	108-89-2	N/R	N/A	N/A	N/A	1-3/8	1-3/8	1-11/16	1-11/16
108-83-2	108-89-2	108-90-2	108-90-2	N/R	134-101	134-105	112-20	1-1/4	1-5/16	1-9/16	1-9/16
108-83-2	108-89-2	108-83-2	108-89-2	N/R	N/A	N/A	N/A	1-3/32	1-3/32	1-1/2	1-1/2
N/S	N/S	N/S	N/S	N/R	N/R	N/R	N/R	1-3/16	N/R	1-11/16	N/R
N/S	N/S	N/S	N/S	N/R	N/R	N/R	N/R	1-9/16	N/R	1-11/16	N/R
N/S	N/S	N/S	N/S	N/R	N/R	N/R	N/R	1-9/16	N/R	1-3/4	N/R
N/A	N/A	N/A	N/A	N/A	N/S	N/S	N/S	1-3/16	N/R	1-11/16	N/R
N/A	N/A	N/A	N/A	N/A	N/S	N/S	N/S	1-9/16	N/R	1-11/16	N/R
N/A	N/A	N/A	N/A	N/A	N/S	N/S	N/S	1-9/16	N/R	1-11/16	N/R
N/A	N/A	N/A	N/A	N/A	N/S	N/S	N/S	1-9/16	N/R	1-11/16	N/R
108-83-2	108-89-2	108-83-2	108-89-2	N/R	N/A	N/A	N/A	1-3/8	1-3/8	1-11/16	1-11/16
108-83-2	108-89-2	108-83-2	108-89-2	N/R	N/A	N/A	N/A	1-3/8	1-3/8	1-11/16	1-11/16
108-83-2	108-89-2	108-90-2	108-90-2	108-27-2	134-103	134-102	N/S	1-3/8	1-7/16	1-11/16	1-11/16
108-83-2	108-89-2	108-90-2	108-90-2	108-27-2	134-103	134-102	N/S	1-3/8	1-7/16	1-11/16	1-11/16
108-83-2	108-89-2	108-83-2	108-89-2	N/R	134-103	134-102	N/S	1-3/8	1-7/16	1-11/16	1-11/16
108-83-2	108-89-2	108-83-2	108-89-2	N/R	134-103	134-102	N/S	1-3/8	1-7/16	1-11/16	1-11/16
-3	-3	-3	-3	-3	N/R	N/R	N/R	1-1/4	1-1/4	1-11/16	1-11/16
-3	-3	-3	-3	-3	N/R	N/R	N/R	1-1/4	1-1/4	1-11/16	1-11/16
-3	-3	-3	-3	-3	N/R	N/R	N/R	1-1/4	1-1/4	1-11/16	1-11/16
-3	-3	-3	-3	-3	N/R	N/R	N/R	1-1/4	1-1/4	1-11/16	1-11/16
-3	-3	-3	-3	-3	N/R	N/R	N/R	1-1/2	1-1/2	1-11/16	1-11/16
-3	-3	-3	-3	-3	N/R	N/R	N/R	1-1/2	1-1/2	1-11/16	1-11/16
-3	-3	-3	-3	-3	N/R	N/R	N/R	1-1/2	1-1/2	1-11/16	1-11/16

Carb Number	Carb Model Number	CFM	Renew Kit	Trick Kit	Needle & Seat	Primary Main Jet	Secondary Main Jet or plate	Primary Metering Block	Secondary Metering Block	Power Valve	Pump Nozzle Size, Primary	Secondary Nozzle Size or Spring Color
R84011-3	4010	750	37-1541	N/A	6-504	122-73	122-75	N/R	N/R	125-65	0.031	Purple
R84012	4010	600	37-1541	N/A	6-504	122-67	122-77	N/R	N/R	125-65	0.026	0.026
R84012-1	4010	600	37-1541	N/A	6-504	122-67	122-77	N/R	N/R	125-65	0.035	0.026
R84012-2	4010	600	37-1541	N/A	6-504	122-67	122-77	N/R	N/R	125-65	0.035	0.026
R84012-3	4010	600	37-1541	N/A	6-504	122-67	122-77	N/R	N/R	125-65	0.031	0.026
R84013	4010	750	37-1541	N/A	6-504	122-79	122-79	N/R	N/R	125-65 (15)	0.026	0.026
R84013-1	4010	750	37-1541	N/A	6-504	122-79	122-79	N/R	N/R	125-65 (15)	0.035	0.026
R84013-2	4010	750	37-1541	N/A	6-504	122-79	122-79	N/R	N/R	125-65 (15)	0.035	0.026
R84013-3	4010	750	37-1541	N/A	6-504	122-75	122-79	N/R	N/R	125-65	0.031	0.026
R84014	4011	650	37-1541	N/A	6-504	122-60	122-66	N/R	N/R	125-65 (15)	0.026	Plain
R84014-1	4011	650	37-1541	N/A	6-504	122-60	122-66	N/R	N/R	125-65 (15)	0.026	Plain
R84014-2	4011	650	37-1541	N/A	6-504	122-60	122-64	N/R	N/R	125-65 (15)	0.026	Plain
R84014-3	4011	650	37-1541	N/A	6-504	122-60	122-64	N/R	N/R	125-65 (15)	0.026	Plain
R84015	4011	800	37-1541	N/A	6-504	122-64	122-90	N/R	N/R	125-65 (15)	0.026	Yellow
R84015-1	4011	800	37-1541	N/A	6-504	122-64	122-90	N/R	N/R	125-65 (15)	0.026	Yellow
R84015-2	4011	800	37-1541	N/A	6-504	122-60	122-90	N/R	N/R	125-65 (15)	0.026	Yellow
R84015-3	4011	800	37-1541	N/A	6-504	122-60	122-90	N/R	N/R	125-65 (15)	0.026	Plain
R84016	4011	650	37-1541	N/A	6-504	122-64	122-64	N/R	N/R	125-65 (15)	0.026	0.026
R84016-1	4011	650	37-1541	N/A	6-504	122-64	122-64	N/R	N/R	125-65 (15)	0.026	0.026
R84016-2	4011	650	37-1541	N/A	6-504	122-60	122-64	N/R	N/R	125-65 (15)	0.026	0.026
R84016-3	4011	650	37-1541	N/A	6-504	122-60	122-64	N/R	N/R	125-65 (15)	0.026	0.026
R84017	4011	800	37-1541	N/A	6-504	122-64	122-90	N/R	N/R	125-65 (15)	0.026	0.026
R84017-1	4011	800	37-1541	N/A	6-504	122-64	122-90	N/R	N/R	125-65 (15)	0.026	0.026
R84017-2	4011	800	37-1541	N/A	6-504	122-60	122-90	N/R	N/R	125-65 (15)	0.026	0.026
R84017-3	4011	800	37-1541	N/A	6-504	122-60	122-90	N/R	N/R	125-65 (15)	0.026	0.026
R84018	4010	750	3-1445	N/A	6-504	122-86	122-86	N/S	N/S	125-65 (15)	0.026	YELLOW
R84020	4010	600	37-1541	N/A	6-504	122-67	122-75	N/R	N/R	125-65	0.026	Purple
R84020-1	4010	600	37-1541	N/A	6-504	122-67	122-75	N/R	N/R	125-65	0.035	Purple
R84020-2	4010	600	37-1541	N/A	6-504	122-67	122-75	N/R	N/R	125-65	0.035	Purple
R84020-3	4010	600	37-1541	N/A	6-504	122-63	122-75	N/R	N/R	125-65	0.035	Purple
R84021	4011	650	37-1541	N/A	6-504	122-60	122-64	N/R	N/R	125-65 (15)	0.026	Plain
R84021-1	4011	650	37-1541	N/A	6-504	122-60	122-64	N/R	N/R	125-65 (15)	0.026	Plain
R84021-2	4011	650	37-1541	N/A	6-504	122-60	122-64	N/R	N/R	125-65 (15)	0.026	Plain
R84021-3	4011	650	37-1541	N/A	6-504	122-60	122-64	N/R	N/R	125-65 (15)	0.026	Plain
R84022	4011	800	3-1447	N/A	6-504	122-64	122-95	N/S	N/S	125-65 (15)	0.026	YELLOW
R84023	4010	600	3-1445	N/A	6-504	122-65	122-75	N/S	N/S	125-65	0.026	YELLOW
R84023-2	4010	600	3-1445	N/A	6-504	122-67	122-75	N/S	N/S	125-65	0.035	YELLOW
R84024	4011	650	3-1447	N/A	6-504	122-60	122-64	N/S	N/S	125-65 (15)	0.026	YELLOW
R84024-1	4011	650	3-1447	N/A	6-504	122-64	122-68	N/S	N/S	125-65 (15)	0.026	YELLOW
R84026	4011	650	3-1447	N/A	6-504	122-64	122-68	N/S	N/S	125-65 (15)	0.026	YELLOW
R84028	4010	750	3-1445	N/A	6-504	122-86	122-90	N/S	N/S	125-65 (15)	0.035	RED
R84035	4010	600	37-1541	N/A	6-504	122-67	122-75	N/R	N/R	125-65	0.035	Purple
R84035-1	4010	600	37-1541	N/A	6-504	122-67	122-75	N/R	N/R	125-65	0.035	Purple
R84035-2	4010	600	37-1541	N/A	6-504	122-63	122-75	N/R	N/R	125-65	0.035	Purple
R84037	4011	650	703-59	N/A	6-504	122-63	122-69	N/S	N/S	125-85 (30)	0.026	YELLOW
R84038	4010	600	3-1445	N/A	6-504	122-67	122-77	N/S	N/S	125-65	0.035	0.035
R84039	4010	750	3-1445	N/A	6-504	122-79	122-79	N/S	N/S	125-65 (15)	0.035	0.035
R84040	4011	650	3-1447	N/A	6-504	122-60	122-64	N/S	N/S	125-65 (15)	0.026	0.026
R84041	4011	800	3-1447	N/A	6-504	122-64	122-90	N/S	N/S	125-65 (15)	0.026	0.026
R84042	4011	650	3-1447	N/A	6-504	122-64	122-68	N/S	N/S	125-65 (15)	0.026	YELLOW
R84044	4010	750	3-1445	N/A	6-504	122-71	122-76	N/S	N/S	125-105 (15)	0.035	RED
R84044-1	4010	750	3-1445	N/A	6-504	122-73	122-76	N/S	N/S	125-105 (15)	0.035	RED
R84046	4010	600	703-53	N/A	6-504	122-69	122-76	N/S	N/S	125-65	0.035	PINK

Primary Bowl Gasket	Primary Metering Block Gasket	Secondary Bowl Gasket	Secondary Metering Block Gasket	Secondary Metering Plate Gasket	Primary Fuel Bowl	Secondary Fuel Bowl	Throttle Body & Shaft Assembly	Venturii Diameter Primary	Venturii Diameter Secondary	Throttle Bore Diameter Primary	Throttle Bore Diameter Secondary
-3	-3	-3	-3	-3	N/R	N/R	N/R	1-1/2	1-1/2	1-11/16	1-11/16
-3	-3	-3	-3	-3	N/R	N/R	N/R	1-1/4	1-1/4	1-11/16	1-11/16
-3	-3	-3	-3	-3	N/R	N/R	N/R	1-1/4	1-1/4	1-11/16	1-11/16
-3	-3	-3	-3	-3	N/R	N/R	N/R	1-1/4	1-1/4	1-11/16	1-11/16
-3	-3	-3	-3	-3	N/R	N/R	N/R	1-1/4	1-1/4	1-11/16	1-11/16
-3	-3	-3	-3	-3	N/R	N/R	N/R	1-1/2	1-1/2	1-11/16	1-11/16
-3	-3	-3	-3	-3	N/R	N/R	N/R	1-1/2	1-1/2	1-11/16	1-11/16
-3	-3	-3	-3	-3	N/R	N/R	N/R	1-1/2	1-1/2	1-11/16	1-11/16
-3	-3	-3	-3	-3	N/R	N/R	N/R	1-1/2	1-1/2	1-11/16	1-11/16
-4	-4	-4	-4	-4	N/R	N/R	N/R	1-5/32	1-3/8	1-3/8	2
-4	-4	-4	-4	-4	N/R	N/R	N/R	1-5/32	1-3/8	1-3/8	2
-4	-4	-4	-4	-4	N/R	N/R	N/R	1-5/32	1-3/8	1-3/8	2
-4	-4	-4	-4	-4	N/R	N/R	N/R	1-5/32	1-3/8	1-3/8	2
-4	-4	-4	-4	-4	N/R	N/R	N/R	1-5/32	1-23/32	1-3/8	2
-4	-4	-4	-4	-4	N/R	N/R	N/R	1-5/32	1-23/32	1-3/8	2
-4	-4	-4	-4	-4	N/R	N/R	N/R	1-5/32	1-23/32	1-3/8	2
-4	-4	-4	-4	-4	N/R	N/R	N/R	1-5/32	1-23/32	1-3/8	2
-4	-4	-4	-4	-4	N/R	N/R	N/R	1-5/32	1-3/8	1-3/8	2
-4	-4	-4	-4	-4	N/R	N/R	N/R	1-5/32	1-3/8	1-3/8	2
-4	-4	-4	-4	-4	N/R	N/R	N/R	1-5/32	1-3/8	1-3/8	2
-4	-4	-4	-4	-4	N/R	N/R	N/R	1-5/32	1-3/8	1-3/8	2
-4	-4	-4	-4	-4	N/R	N/R	N/R	1-5/32	1-23/32	1-3/8	2
-4	-4	-4	-4	-4	N/R	N/R	N/R	1-5/32	1-23/32	1-3/8	2
-4	-4	-4	-4	-4	N/R	N/R	N/R	1-5/32	1-23/32	1-3/8	2
-4	-4	-4	-4	-4	N/R	N/R	N/R	1-5/32	1-23/32	1-3/8	2
-3	-3	-3	-3	-3	N/S	N/S	N/S	1-1/2	1-1/2	1-11/16	1-11/16
-3	-3	-3	-3	-3	N/R	N/R	N/R	1-1/4	1-1/4	1-11/16	1-11/16
-3	-3	-3	-3	-3	N/R	N/R	N/R	1-1/4	1-1/4	1-11/16	1-11/16
-3	-3	-3	-3	-3	N/R	N/R	N/R	1-1/4	1-1/4	1-11/16	1-11/16
-3	-3	-3	-3	-3	N/R	N/R	N/R	1-1/4	1-1/4	1-11/16	1-11/16
-4	-4	-4	-4	-4	N'R	N/R	N/R	1-5/32	1-3/8	1-3/8	2
-4	-4	-4	-4	-4	N/R	N/R	N/R	1-5/32	1-3/8	1-3/8	2
-4	-4	-4	-4	-4	N/R	N/R	N/R	1-5/32	1-3/8	1-3/8	2
-4	-4	-4	-4	-4	N/R	N/R	N/R	1-5/32	1-3/8	1-3/8	2
-4	-4	-4	-4	-4	N/S	N/S	N/S	1-5/32	1-23/32	1-3/8	2
-3	-3	-3	-3	-3	V/S	N/S	N/S	1-1/4	1-1/4	1-11/16	1-11/16
-3	-3	-3	-3	-3	N/S	N/S	N/S	1-1/4	1-1/4	1-11/16	1-11/16
-4	-4	-4	-4	-4	N/S	N/S	N/S	1-5/32	1-3/8	1-3/8	2
-4	-4	-4	-4	-4	N/S	N/S	N/S	1-5/32	1-3/8	1-3/8	2
-4	-4	-4	-4	-4	N/S	N/S	N/S	1-5/32	1-3/8	1-3/8	2
-3	-3	-3	-3	-2	N/S	N/S	N/S	1-1/2	1-1/2	1-11/16	1-11/16
-3	-3	-3	-3	-3	N/R	N/R	N/R	1-1/4	1-1/4	1-11/16	1-11/16
-3	-3	-3	-3	-3	N/R	N/R	N/R	1-1/4	1-1/4	1-11/16	1-11/16
-3	-3	-3	-3	-3	N/R	N/R	N/R	1-1/4	1-1/4	1-11/16	1-11/16
-4	-4	-4	-4	-4	N/S	N/S	N/S	1-5/32	1-3/8	1-3/8	2
-3	-3	-3	-3	-3	N/S	N/S	N/S	1-1/4	1-1/4	1-11/16	1-11/16
-3	-3	-3	-3	-3	N/S	N/S	N/S	1-1/2	1-1/2	1-11/16	1-11/16
-4	-4	-4	-4	-4	N/S	N/S	N/S	1-5/32	1-3/8	1-3/8	2
-4	-4	-4	-4	-4	N/S	N/S	N/S	1-5/32	1-28/32	1-3/8	2
-4	-4	-4	-4	-4	N/S	N/S	N/S	1-5/32	1-3/8	1-3/8	2
-3	-3	-3	-3	-3	N/S	N/S	N/S	1-1/2	1-1/2	1-11/16	1-11/16
-3	-3	-3	-3	-3	N/S	N/S	N/S	1-1/2	1-1/2	1-11/16	1-11/16
-3	-3	-3	-3	-3	N/S	N/S	N/S	1-1/4	1-1/4	1-11/16	1-11/16

Carb Number	Carb Model Number	CFM	Renew Kit	Trick Kit	Needle & Seat	Primary Main Jet	Secondary Main Jet or plate	Primary Metering Block	Secondary Metering Block	Power Valve	Pump Nozzle Size, Primary	Secondary Nozzle Size or Spring Color
R84047	4010	750	37-1541	N/A	6-504	122-75	122-75	N/R	N/R	125-65 (15)	0.035	Black
R84047-1	4010	750	37-1541	N/A	6-504	122-73	122-75	N/R	N/R	125-65 (15)	0.031	Purple
R84050	4011	650	703-60	N/A	6-504	122-58	122-84	N/S	N/S	125-85 (12)	0.026	0.026
R84412	2300	500	37-474	37-933	6-504	122-73	N/R	134-137	N/R	125-50	0.028	N/R
R84776	4150	600	37-485	37-933	6-504	122-66	122-73	N/A	N/A	125-65	0.028	0.032
R84777	4150	650	37-485	37-933	6-504	122-67	122-73	134-150	N/A	125-65	0.028	0.028
R84778	4150	700	37-485	37-933	6-504	122-69	122-78	N/S	N/S	125-65	0.028	0.031
R84779	4150	750	37-485	37-933	6-504	122-70	122-73	N/A	N/S	125-65	0.028	0.031
R84780	4150	800	37-485	37-933	6-504	122-71	122-85	N/A	N/S	125-65	0.031	0.031
R84781	4150	850	37-485	37-933	6-504	122-80	122-78	N/S	N/S	125-65	0.031	0.031
R86670AA	AL 4150	670	37-935	37-933	6-504	122-63	122-73	N/S	N/S	125-65	0.031	Plain
R86770AA	AL 4150	770	37-935	37-933	6-504	122-73	122-80	N/S	N/S	125-65	0.031	Plain
R87448	2300	350	37-1536	37-933	6-504	122-61	N/A	134-203	N/R	125-85	0.031	N/R
R89834	4160	600	37-720	37-933	6-506	122-68	134-39	N/S	134-39	125-65	0.031	Black
R90470	4150	470	37-936	N/A	6-513	122-57	122-57	N/A	N/A	125-25	0.035	Black
R90670	4150	670	N/A	N/A	6-513 (2)	122-68	122-89	N/S	N/S	125-65	0.028	Black
R90670-1	4150	670	37-936	N/A	6-513	122-68	122-89	N/S	N/S	125-65	0.028	Black
R90670-2	4150	670	37-936	N/A	6-513	122-68	122-89	N/S	N/S	125-65	0.028	Black
R90770	4150	770	37-936	N/A	6-513	125-74	122-99	N/A	N/A	125-25	0.035	Black

Primary Bowl Gasket	Primary Metering Block Gasket	Secondary Bowl Gasket	Secondary Metering Block Gasket	Secondary Metering Plate Gasket	Primary Fuel Bowl	Secondary Fuel Bowl	Throttle Body & Shaft Assembly	Venturii Diameter Primary	Venturii Diameter Secondary	Throttle Bore Diameter Primary	Throttle Bore Diameter Secondary
-3	-3	-3	-3	-3	N/R	N/R	N/R	1-1/2	1-1/2	1-11/16	1-11/16
-3	-3	-3	-3	-3	N/R	N/R	N/R	1-1/2	1-1/2	1-11/16	1-11/16
-4	-4	-4	-4	-4	N/S	N/S	N/S	1-5/32	1-3/8	1-3/8	2
108-83-2	108-89-2	N/R	N/R	N/R	134-103	N/R	112-2	1-3/8	N/R	1-11/16	N/R
108-83-2	108-89-2	108-83-2	108-89-2	N/R	134-103	134-104	N/A	1-1/4	1-5/16	1-9/16	1-9/16
108-83-2	108-89-2	108-83-2	108-89-2	N/R	134-103	134-104	112-17	1-1/4	1-5/16	1-11/16	1-11/16
108-83-2	108-89-2	108-83-2	108-89-2	N/R	134-103	134-104	N/A	1-5/16	1-3/8	1-11/16	1-11/16
108-83-2	108-89-2	108-83-2	108-89-2	N/R	134-103	134-104	N/S	1-3/8	1-3/8	1-11/16	1-11/16
108-83-2	108-89-2	108-83-2	108-89-2	N/R	134-103	134-104	N/A	1-3/8	1-7/16	1-11/16	1-11/16
108-83-2	108-89-2	108-83-2	108-89-2	N/R	134-103	134-104	N/S	1-9/16	1-9/16	1-3/4	1-3/4
108-83-2	108-89-2	108-83-2	108-89-2	N/A	N/S	N/S	N/S	1-1/4	1-5/16	1-11/16	1-11/16
108-83-2	108-89-2	108-83-2	108-89-2	N/A	N/S	N/S	N/S	1-3/8	1-3/8	1-11/16	1-11/16
108-83-2	108-89-2	N/R	N/R	N/R	134-103	N/R	N/A	1-3/16	N/R	1-1/2	N/R
108-83-2	108-91-2	108-90-2	108-90-2	108-27-2	N/A	134-105	N/S	1-1/4	1-5/16	1-9/16	1-9/16
108-83-2	108-89-2	108-83-2	108-89-2	N/R	N/A	N/A	N/A	1-3/32	1-3/32	1-1/2	1-1/2
108-83-2	108-89-2	108-83-2	108-89-2	N/R	N/S	N/S	N/S	1-1/4	1-5/16	1-9/16	1-9/16
108-83-2	108-89-2	108-83-2	108-89-2	N/R	N/S	N/S	N/S	1-1/4	1-5/16	1-9/16	1-9/16
108-83-2	108-89-2	108-83-2	108-89-2	N/R	N/S	N/S	N/S	1-1/4	1-5/16	1-9/16	1-9/16
108-83-2	108-89-2	108-83-2	108-89-2	N/R	N/A	N/A	N/A	1-3/8	1-7/16	1-11/16	1-11/16

SOURCE GUIDE

AED Performance
2530 Willis Rd.
Richmond, VA 23237
804-271-9107
aedperformance.com

BLP Racing Products
1015 W. Church St.
Orlando, FL 32805
407-422-0394
blp.com

Braswell Competition Carburetors
7671 N. Business Park Dr.
Tucson, AZ 85743-9622
520-579-9177
braswell.com

CRC Industries
885 Louis Dr.
Warminster, PA 18974
800-556-5074
crcindustries.com

Goodson Tools & Supplies
156 Galewski Dr.
Winona, MN 55987
800-533-8010
goodson.com

Holley Performance Products
1801 Russellville Rd.
Bowling Green, KY 42101
866-464-6553
holley.com

Lisle
P.O. Box 89
Clarinda, IA 51632
712-542-5101
lislecorp.com

Lista International
106 Lowland St.
Holliston, MA 01746
800-722-3020
listaintl.com

MAC Tools
505 N. Cleveland Ave.
Westerville, OH 43082
800-622-8665
mactools.com

Moroso Performance Products
80 Carter Dr.
Guilford, CT 06437
203-453-6571
moroso.com

Quick Fuel Technology
129 Dishman Ln.
Bowling Green, KY 42101
270-793-0900
quickfueltechnology.com

Summit Racing
800-230-3030
summitracing.com

Willy's Carburetor & Dyno Shop
1000 N. Market St.
Mt. Carmel, IL 62863
618-262-8021
willyscarb.com

Holley 4150/4160

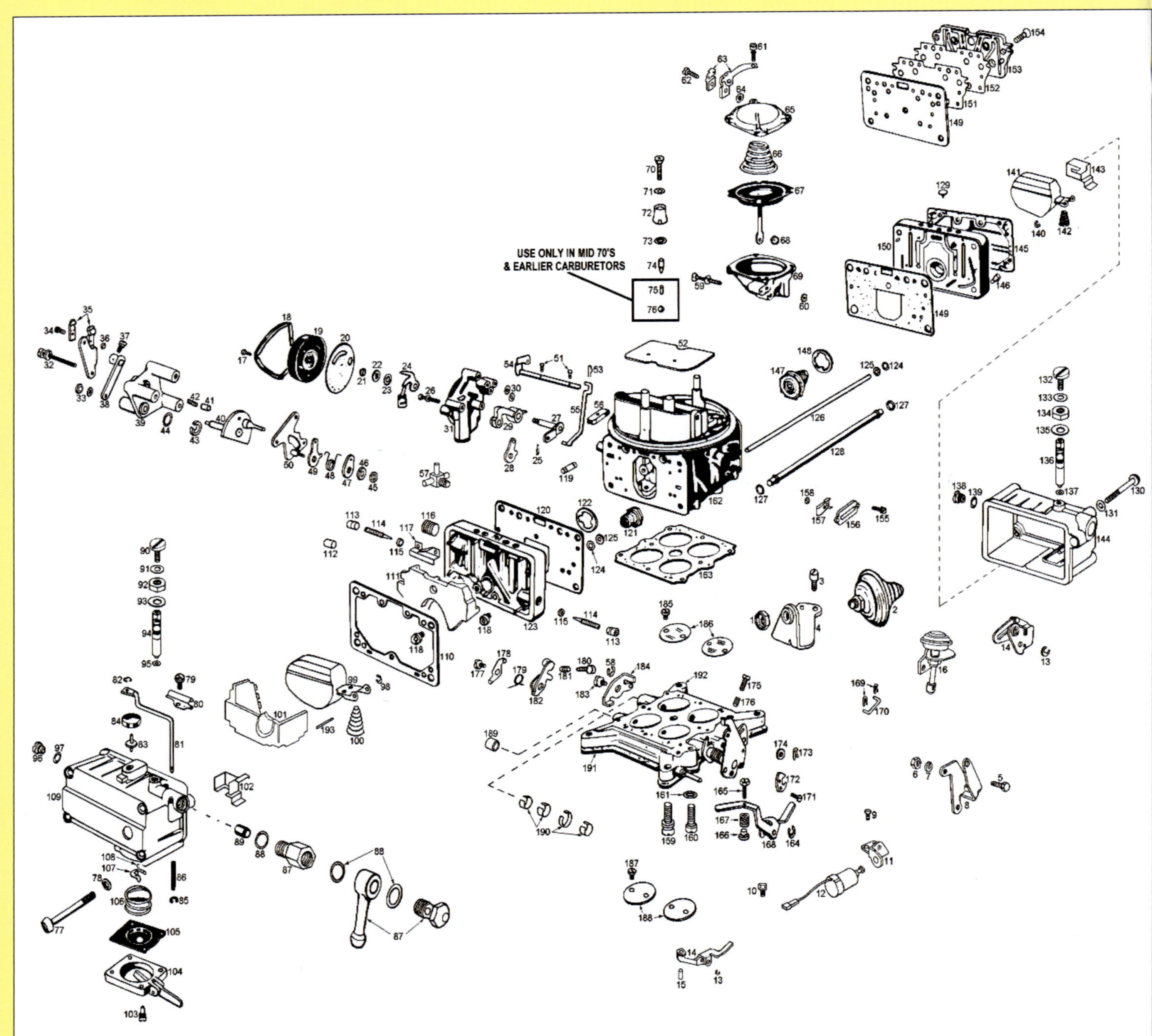

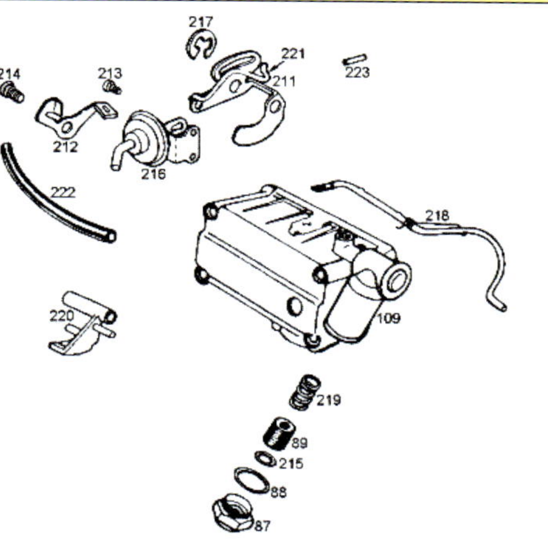

End Inlet Moraine Filter Bowl

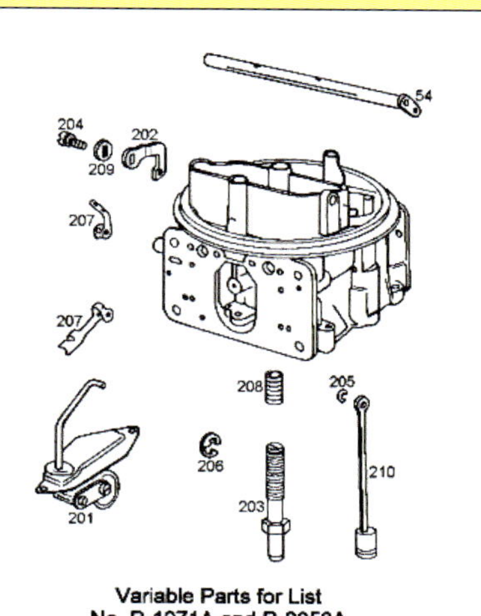

Variable Parts for List
No. R-1971A and R-2052A

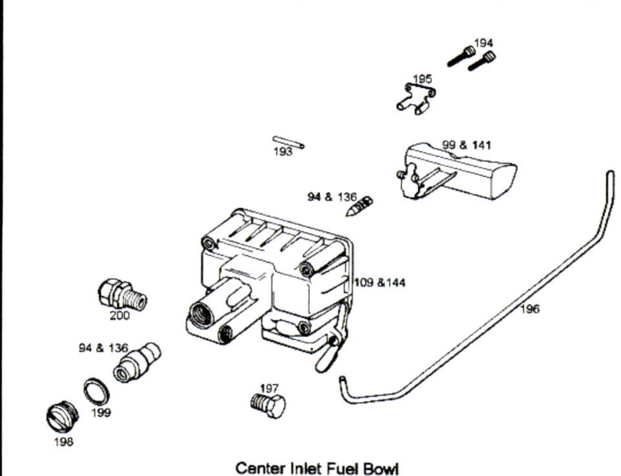

Center Inlet Fuel Bowl

Ref. No.	Description	Ref. No.	Description
1	Dashpot Nut	2	Dashpot Assembly
3	Dashpot Bracket Screw	4	Dashpot Bracket
5	Throttle Lever Extension Screw	6	Throttle Lever Extension Nut
7	Lockwasher Extension Nut	8	Throttle Lever Extension
9	Solenoid Bracket Screw	10	Solenoid Bracket Screw
11	Solenoid Bracket	12	Solenoid Assembly
13	Cam Follower Retainer	14	Cam Follower Lever
15	Pump Operating Lever Screw Sleeve	16	Modulator Assembly
17	Thermostat Housing Clamp Screw	18	Thermostat Housing Clamp
19	Thermostat Housing Assembly	20	Thermostat Housing Gasket
21	Choke Thermostat Shaft Nut	22	Shaft Nut Lockwasher
23	Thermostat Lever Spacer	24	Thermostat Lever, Link & Piston Assembly
25	Choke Rod Retainer	26	Choke Housing Screw
27	Choke Housing Lever & Shaft Assembly	28	Choke Thermostat Lever
29	Fast Idle Cam Assembly	30	Choke Housing Gasket
31	Choke Housing Assembly	32	Fast Idle Cam Plate Screw
33	Choke Control Lever Nut & Lockwasher	34	Wire Bracket Clamp Screw
35	Choke Control Wire Bracket Clamp	36	Wire Bracket Clamp Screw Nut
37	Choke Lever Assembly Swivel Screw	38	Choke lever & Swivel Assembly
39	Fast Idle Cam Plate	40	Fast Idle Cam and Shaft Assembly
41	Fast idle Cam Plunger	42	Fast Idle Cam Plunger Spring
43	Choke Operating Lever Spring Washer	44	Choke Operating lever Washer
45	Back-up Plate Stud Nut	46	Stud Nut Lockwasher
47	Choke Spring Washer	48	Choke Spring
49	Choke Rod & Lever Bushing Assembly	50	Back-up Plate & Stud Assembly
51	Choke Plate Screw	52	Choke Plate
53	Choke rod Retainer	54	Choke lever & Shaft Assembly
55	Choke Rod	56	Choke Rod Seal
57	Four-way Hose Connector	58	Secondary Diaphragm Link Retainer
59	Secondary Diaphragm Housing Screw	60	Secondary Diaphragm Housing Gasket
61	Secondary Diaphragm Cover Screw	62	Wire Bracket & Clamp Screw
63	Wire Bracket Clamp	64	Clamp Screw Nut
65	Secondary Diaphragm Cover	66	Secondary Diaphragm Spring
67	Secondary Diaphragm Housing Check	68	Secondary Diaphragm Housing Check Ball
69	Secondary Diaphragm Housing	70	Pump Discharge Nozzle Screw
71	Pump Discharge Nozzle Screw Gasket	72	Pump Discharge Nozzle
73	Pump Discharge Nozzle Gasket	74	Pump Discharge Needle Valve
75	Pump Check Ball Weight	76	Pump Discharge Nozzle Check Ball
77	Primary Fuel Bowl Screw	78	Primary Fuel Bowl Screw Gasket
79	Air Vent Clamp Screw	80	Air Vent Rod Clamp
81	Air Vent rod	82	Air Vent Valve Retainer
83	Air Vent valve	84	Air Vent Cap
85	Air Vent Rod Spring Retainer	86	Air Vent Rod Spring

Ref. No.	Description	Ref. No.	Description
87	Fuel Inlet Fitting	88	Fuel Inlet Fitting Gasket
89	Fuel Inlet Filter	90	Primary Fuel Valve Seat Lockscrew
91	Fuel Valve Seat Lockscrew Gasket	92	Fuel Valve Seat Adjustment Nut
93	Fuel Valve Adjustment Nut Gasket	94	Primary Fuel Inlet & Valve Seat Assembly
95	Fuel Valve Seat O-Ring Seal	96	Fuel Level Check Plug
97	Fuel Level Check Plug Gasket	98	Float Retainer
99	Primary Float Assembly	100	Primary Float spring
101	Primary Fuel Bowl Filler	102	Primary Baffle Plate
103	Pump Diaphragm Cover Screw & Lockwasher	104	Primary Pump Diaphragm Cover Assembly
105	Primary Pump Diaphragm Assembly	106	Primary Diaphragm Return Spring
107	Primary Check Ball Retainer	108	Primary Check Ball or Check Valve
109	Primary Fuel Bowl Assembly	110	Primary Fuel Bowl Gasket
111	Primary Metering Body Filler	112	Vacuum Tube Plug
113	Idle Limiter Cap	114	Idle Adjustment Needle
115	Idle Adjustment Needle Seal	116	Spark Hole Plug
117	Metering Body Vent Baffle	118	Primary Metering jet
119	Tube & O-Ring Assembly	120	Primary Metering Body Gasket
121	Primary Power Valve Assembly	122	Primary Power Valve Gasket
123	Primary Metering Body Assembly	124	Balance Tube Washer
125	Balance Tube O-Ring Seal	126	Balance Tube
127	Fuel Line Tube O-Ring Seal	128	Fuel Line Tube
129	Secondary Bowl Vent Valve	130	Secondary Fuel Bowl Screw
131	Secondary Fuel Bowl Screw Gasket	132	Secondary Fuel Valve Seat Lock Screw
133	Fuel Valve Seat Lock Screw Gasket	134	Secondary Fuel Valve Seat Adjustment Nut
135	Fuel Valve Seat Adjustment Nut Gasket	136	Secondary Fuel Inlet Valve & Seat Assembly
137	Fuel Valve Seat O-Ring Seal	138	Secondary Fuel Level Check Plug
139	Fuel Level Check Plug Gasket	140	Float Retainer
141	Secondary Float Assembly	142	Secondary Float Spring
143	Secondary Baffle Plate	144	Secondary Fuel Bowl Assembly
145	Secondary Fuel Bowl Gasket	146	Secondary Metering Jet
147	Secondary Power Valve Assembly	148	Secondary Power Valve Gasket
149	Secondary Metering Body Gasket	150	Secondary Metering Body Assembly
151	Secondary Metering Body Plate	152	Secondary Metering Body Plate Gasket
153	Main Metering Body & Plugs Assembly - Secondary	154	Secondary Metering Body Screw
155	H.I.C. Cover Screw	156	H.I.C. Cover
157	Hot Idle Compensator Assembly	158	H.I.C. Seal
159	Throttle Body Screw & Lockwasher	160	Throttle Body Screw
161	Throttle Body Screw Gasket	162	Main Body Assembly
163	Throttle Body Gasket	164	Pump Operating Lever Retainer
165	Pump Lever Adjustment Screw	166	Pump Lever Adjustment Screw Fitting
167	Pump Lever Adjustment Screw Spring	168	Pump Lever
169	Throttle Connector Rod Cotter Pin	170	Throttle Connector Rod
171	Pump Cam Lock Screw	172	Pump Cam
173	Throttle Connector Pin Retainer	174	Throttle Connector Pin Washer
175	Throttle Stop Screw	176	Throttle Stop Screw Spring
177	Fast Idle Cam lever Screw & Lockwasher	178	Fast Idle Cam Pick-up Lever
179	Fast Idle Cam Lever Spring	180	Fast Idle Cam Lever Adjustment Screw
181	Fast Idle Cam Lever Adjustment Screw Spring	182	Fast Idle Cam Lever
183	Secondary Diaphragm Lever Assembly Screw	184	Secondary Diaphragm Lever Assembly
185	Secondary Throttle Plate Screw	186	Secondary Throttle Plate
187	Primary Throttle Plate Screw	188	Primary Throttle Plate
189	Primary Throttle Shaft Bearing (solid)	190	Throttle Shaft Bearing (Ribbon) Primary and Secondary
191	Flange Gasket	192	Throttle Body Assembly
193	Float Lever Shaft	194	Float Shaft Bracket Screw & Lockwasher
195	Float Shaft Retainer Bracket	196	Fuel Transfer Tube
197	Fuel Bowl Drain Plug	198	Fuel Bowl Plug
199	Fuel Bowl Plug Gasket	200	Fuel Transfer Tube Fitting Assembly
201	Choke Assembly - Complete (divorced)	202	Choke Shaft Lever
203	Idle By-pass Adjustment Screw	204	Choke Lever Screw and Lockwasher
205	Choke Piston Link Retainer	206	Fast Idle Cam Retainer
207	Choke Rod Clevis Clip	208	Idle By-pass Adjustment Screw Spring
209	Choke Piston Lever Spacer	210	Choke Piston and Link Assembly
211	Choke Control Lever	212	Fast Idle Cam Lever
213	Choke Diaphragm, Assy., Brkt. Screw, & Lockwasher	214	Fast Idle Cam lever Screw and Lockwasher
215	Fuel Inlet Filter Gasket	216	Choke Diaphragm Assembly - Complete
217	Choke Control Lever Retainer	218	Air Vent Rod Spring
219	Fuel Inlet Filter Spring	220	Fast Idle Cam Assembly
221	Choke Diaphragm Assembly Link	222	Choke Vacuum Tube